Introduction to Chemical Graph Theory

T0177403

Discrete Mathematics and Its Applications

Series Editors
Miklos Bona
Donald L. Kreher
Douglas West
Patrice Ossona de Mendez

Graph Polynomials
Yongtang Shi, Matthias Dehmer, Xueliang Li, and Ivan Gutman

Introduction to Combinatorics, Second Edition
W. D. Wallis and J. C. George

Representation Theory of Symmetric Groups
Pierre-Loïc Méliot

Advanced Number Theory with Applications
Richard A. Mollin

A Multidisciplinary Introduction to Information Security
Stig F. Mjølsnes

Combinatorics of Compositions and Words
Silvia Heubach and Toufik Mansour

Handbook of Linear Algebra, Second Edition
Leslie Hogben

Combinatorics, Second Edition
Nicholas A. Loehr

Handbook of Discrete and Computational Geometry, Third Edition
C. Toth, Jacob E. Goodman and Joseph O'Rourke

Handbook of Discrete and Combinatorial Mathematics, Second Edition
Kenneth H. Rosen

Crossing Numbers of Graphs
Marcus Schaefer

Graph Searching Games and Probabilistic Methods
Anthony Bonato and Paweł Prałat

Handbook of Geometric Constraint Systems Principles
Meera Sitharam, Audrey St. John, and Jessica Sidman

Introduction to Chemical Graph Theory
Stephan Wagner and Hua Wang

https://www.crcpress.com/Discrete-Mathematics-and-Its-Applications/book-series/
CHDISMTHAPP?page=1&order=dtitle&size=12&view=list&status=published,forthcoming

DISCRETE MATHEMATICS AND ITS APPLICATIONS

Introduction to Chemical Graph Theory

Stephan Wagner
Hua Wang

CRC Press
Taylor & Francis Group
Boca Raton London New York

CRC Press is an imprint of the
Taylor & Francis Group, an **informa** business

A CHAPMAN & HALL BOOK

CRC Press
Taylor & Francis Group
6000 Broken Sound Parkway NW, Suite 300
Boca Raton, FL 33487-2742

First issued in paperback 2022

Version Date: 20180817

ISBN 13: 978-1-03-247603-2 (pbk)
ISBN 13: 978-1-138-32508-1 (hbk)

DOI: 10.1201/9780429450532

Library of Congress Cataloging-in-Publication Data

Names: Wagner, Stephan, 1982- author. | Wang, Hua (Mathematics professor), author.
Title: Introduction to chemical graph theory / Stephan Wagner and Hua Wang.
Other titles: Chemical graph theory
Description: Boca Raton : CRC Press, Taylor & Francis Group, 2018. | Includes index.
Identifiers: LCCN 2018019473 | ISBN 9781138325081
Subjects: LCSH: Chemistry--Mathematics. | Graph theory.
Classification: LCC QD39.3.G73 W34 2018 | DDC 540.1/5115--dc23
LC record available at https://lccn.loc.gov/2018019473

Visit the Taylor & Francis Web site at
http://www.taylorandfrancis.com

and the CRC Press Web site at
http://www.crcpress.com

Contents

Preface

Chemical graph theory is an interdisciplinary field where the molecular structure of a chemical compound is analyzed as a graph, and where related mathematical questions are investigated through graph theoretical and computational techniques. The rapid development in this field in the last few decades has presented us with many innovative and unique concepts and tools in such studies. This book intends to introduce some of the most commonly used mathematical approaches in chemical graph theory.

One of the most important ideas employed in chemical graph theory is that of so-called chemical indices. This is to associate a numerical value with a graph structure that often has some kind of correlation with corresponding chemicals' properties. For exactly this reason, these chemical indices are generally considered descriptors of chemical structures. The investigation of such a chemical index from a graph-theoretical point of view typically involves studying its behavior in various classes of graphs, especially minima and maxima as well as upper and lower bounds in terms of different graph parameters.

In recent years numerous chemical indices have been proposed and as Ivan Gutman pointed out, "we have far too many descriptors, and there seems to lack a firm criterion to stop or slow down their proliferation". Indeed, to extensively study each chemical index is virtually impossible and likely not informative. In this book, we chose to demonstrate some of the most common questions and ideas through several major classes of chemical indices and their representatives. These important classes of chemical indices include distance-based indices, degree-based indices, indices based on counting specific sets, and indices associated with matrices and their spectra.

After a brief introduction of related graph theoretical terminologies, we start with the Wiener index as a representative of distance-based indices. We then look at the Randić index and degree-based indices in general. As representatives of indices defined by "counting", we explore the number of independent vertex sets and the number of independent edge sets (matchings), which are known as the Merrifield-Simmons index and the Hosoya index, respectively. Lastly, graph spectra and graph energy, defined through matrices associated with graphs, are discussed.

By no means are we attempting to be comprehensive when we discuss the content related to each of these classes of indices. Instead, we picked, to our best knowledge, some of the most commonly discussed problems and their solutions. This choice is inherently subjective and reflects, at least in part, the authors' personal interests. Through these selected topics, we hope to illustrate at least some of the useful techniques that are used to answer questions in chemical graph theory.

1

Preliminaries

1.1 Basic graph notations

As can be found in any graph theory textbook, a graph G consists of a pair $(V(G), E(G))$, where $V(G)$ (or simply V when there is no ambiguity) is the set of vertices and $E(G)$ (or simply E when there is no ambiguity) is the set of edges. The cardinalities of $V(G)$ and $E(G)$ are often called the *order* and *size* of a graph G. In practice we often use $|G|$ to denote $|V(G)|$ (but not $|E(G)|$).

Each edge in $E(G)$ connects two vertices from $V(G)$, called the ends or endpoints of this edge. For an edge e with endpoints u and v, we say that u and v are *adjacent* to each other, and that u and v are *incident* to e. The set of all vertices adjacent to a specific vertex v is called the (open) *neighborhood* of v and denoted by $N_G(v)$ (or just $N(v)$ for short). The set $N[v] = N(v) \cup \{v\}$ that also includes v itself is called the *closed neighborhood*.

The number of edges that a vertex v is incident to in G is called the *degree* of v, denoted by $\deg_G(v)$ or simply $\deg(v)$. Note that $\deg(v) = |N(v)|$. A *path* in G is a sequence of distinct vertices in G such that each vertex is adjacent to the next. The length of a path is the number of edges on this path. The *distance* between two vertices $u, v \in V(G)$, denoted by $d_G(u, v)$ or simply $d(u, v)$, is the number of edges on the shortest path between them in G. A vertex of degree 1 is called a *pendant vertex* or a *leaf*, a vertex of degree at least 2 is called an *internal vertex*, and a vertex of degree at least 3 is called a *branching vertex*. See Figure 1.1 for an example of these definitions.

FIGURE 1.1
An example where v is a leaf, u is a branching vertex of degree 5, and $d(u, v) = 4$.

A cycle is a path with an additional edge connecting the starting and ending vertices. A graph without a cycle is called *acyclic*. A graph is *connected* if there is a path between every pair of vertices. A connected acyclic graph is called a *tree*; see the next section. For a general graph (not necessarily connected), the *connected components* are the maximal subgraphs that are connected.

In chemical graph theory, the molecular structure of a compound is often presented with a graph, where the atoms are represented by vertices and bonds are represented by edges. Note that the difference between double bonds and single bonds is often ignored. Consequently, there are usually no multiple edges (no two vertices serve as the endpoints of more than one edge). The hydrogen atoms will automatically be leaves in such a representation as the valence of a hydrogen atom is 1, corresponding to its vertex degree. Hence, we usually remove the vertices corresponding to hydrogens. As a result we have what is known as the *molecular graph*. See Figure 1.2 for an example. Such representations appeared as early as 1874 [13], as pointed out in [7].

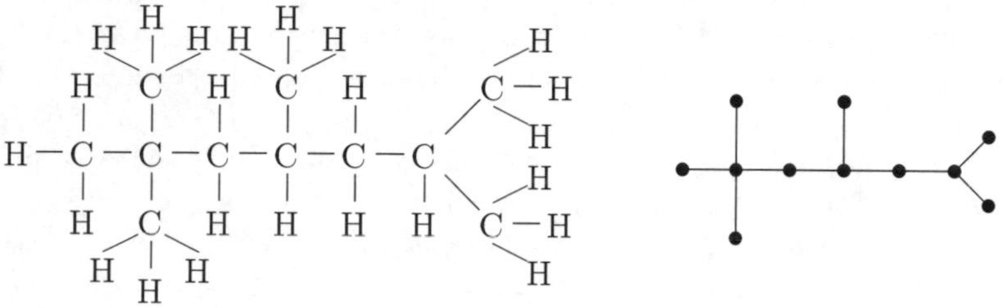

FIGURE 1.2
Structural formula for 2,2,4,6-tetramethylheptane (on the left) and its corresponding molecular graph (on the right).

Most of the chemical compounds under consideration are carbon-based. The vertex degrees are no more than 4, corresponding to the valence of carbon. Often one uses the term *chemical graphs* to refer to the graphs whose vertex degrees are bounded above by 4.

1.2   Special types of graphs

There are a number of special graphs that occur very frequently in graph theory, and in particular throughout this book. Perhaps the simplest examples are the complete and edgeless graphs. A *complete graph* K_n on n vertices is one where every two vertices are adjacent to each other. By contrast, an *edgeless graph* E_n is a graph on n vertices without any edges.

A graph is called *bipartite* if its vertices can be divided into two sets A and B (called the *partite sets*) such that all edges have one end in A and one end in B. More generally, a *k-partite* graph is a graph whose vertices can be divided into sets $A_1, A_2, \ldots, A_k$ such that there is no edge between two vertices that belong to the same set A_i. The smallest integer k such that a graph G is k-partite is also known as the *chromatic number* of G: the smallest number of colors needed to color all vertices in such a way that no two adjacent vertices have the same color. A *complete bipartite graph* $K_{a,b}$ consists of two sets of vertices A and B with $|A| = a$ and $|B| = b$ and all possible edges between A and B. See Figure 1.3 for examples.

FIGURE 1.3
The complete graph K_4 and the complete bipartite graph $K_{2,3}$.

1.3 Trees

As mentioned before, a tree is a connected acyclic graph. A graph that is only acyclic, but not necessarily connected, is also called a *forest* (it can be seen as a union of trees). Since many compounds have acyclic molecular structures, trees have been an important class of graphs in the study of chemical graph theory.

It is well known (and easy to prove by induction) that for a tree T we have $|E(T)| = |V(T)| - 1$, i.e., the number of edges is one less than the number of vertices. It will often be important that every tree with at least two vertices has at least two leaves. It is also worth mentioning that all trees are bipartite.

The notion of *branches* will frequently play a role: if v is a vertex of a tree T, then the branches of v are the connected components of the graph that results from removing v.

A tree is *rooted* if there is a specified vertex designated as the *root*. In a rooted tree the *height* of a vertex v, denoted by $h(v)$, is the distance between v and the root. The height of the tree T, denoted by $h(T)$, is the largest height of any vertex. For two vertices u and v that are adjacent to each other, if $h(u) < h(v)$, then u is called the *parent* of v and v is a *child* of u. More generally, if u is on the path connecting the root and a vertex v, then v is a *descendant* of u and u is an *ancestor* of v.

Two trees of order n will occur particularly frequently: the *path* P_n is the only tree with only two leaves, and the *star* S_n is the only tree with $n - 1$ leaves (Figure 1.4). Intuitively, the path is the most "stretched out" among all trees of the same order, and the star is the most "compact" among all trees of the same order. In fact, the path and the star turn out to be the extremal structures in the studies of many topics in chemical graph theory.

FIGURE 1.4
A star (on the left) and a path (on the right).

Sometimes acyclic structures are needed that are "compact" on one end and "stretched out" on the other, resulting in the so-called *comet* formed from appending multiple pendant edges to one end of a path (Figure 1.5).

FIGURE 1.5
An example of a comet.

When trees with specific constraints are considered, many problems become much more complicated and various special trees need to be defined. An example of this kind is the class of *caterpillars*: a caterpillar is a tree with the property that a path remains when all leaves are removed; see Figure 1.6 for an example.

FIGURE 1.6
An example of a caterpillar.

We observe that a star has only one internal (non-leaf) vertex. A slightly more general notion is that of a *starlike tree*, a tree with only one branching vertex: given a sequence $(l_1, l_2, \ldots, l_m)$ of positive integers, the starlike tree $S(l_1, l_2, \ldots, l_m)$ is the tree with exactly one vertex of degree ≥ 3 formed by identifying one end of each of m paths of length $l_1, l_2, \ldots, l_m$, respectively. See Figure 1.7 for an example.

Similarly, note that caterpillars are characterized by the property that all non-leaves lie on a single path. Relaxing this condition slightly leads naturally to the notion of *quasi-caterpillars*. A quasi-caterpillar is a tree with the property that all its branching vertices lie on a path; see Figure 1.8.

Starlike trees, caterpillars and quasi-caterpillars will occur repeatedly as extremal structures in Chapter 2.

FIGURE 1.7
An example of a starlike tree.

FIGURE 1.8
A quasi-caterpillar.

1.4 Degrees in graphs

As we have already mentioned, vertex degrees play a key role in molecular graphs because of their correlation with the valences of atoms. The most important and fundamental concept based on vertex degrees is probably the *degree sequence*: the non-increasing sequence of vertex degrees of a graph. A degree sequence is called *graphical* if there exists a *simple graph* (a graph with no multiple edges or loops) that realizes this degree sequence.

The following well known identity, sometimes called the "handshake lemma", relates the degree sequence to the number of vertices: if $d_1, d_2, \ldots, d_n$ are the vertex degrees of a graph, and m is the number of edges, then

$$\sum_{i=1}^{n} d_i = 2m. \tag{1.1}$$

This identity is based on the observation that each edge is counted twice (once for each end) in the degree sum.

In particular, we see that not all sequences of non-negative integers are graphical: since the degree sum equals $2m$, it must be even. Consequently the number of odd degrees in a graphical degree sequence must be even. This condition on its own, however, is still insufficient, as can be seen from examples like the sequence $(3, 1)$. The following sufficient and necessary condition is due to Erdős and Gallai [26]:

Theorem 1.4.1 *A sequence* $d_1 \geq d_2 \geq \ldots \geq d_n$ *is graphical if and only if* $\sum_{i=1}^{n} d_i$ *is even and*

$$\sum_{i=1}^{k} d_i \leq k(k-1) + \sum_{i=k+1}^{n} \min\{d_i, k\}$$

for all k.

Much useful information about the graph can be extracted from the degree sequence. Let G be a graph with degree sequence $\pi = (d_1, d_2, \ldots, d_n)$ with

$$\sum_{i=1}^{n} d_i = 2(n + c - 1) \text{ and } d_1 \geq d_2 \geq c + 1 \geq 1.$$

The *cyclomatic number* of G is

$$|E(G)| - |V(G)| + 1 = \frac{1}{2}\sum_{i=1}^{n} d_i - n + 1 = c,$$

which corresponds to the number of independent (in the sense of linear independence in the so-called cycle space) cycles in G. As special cases, connected graphs with the cyclomatic number $c = 0$ are trees, connected graphs with cyclomatic number $c = 1$ are called *unicyclic* graphs, and connected graphs with cyclomatic number $c = 2$ are called *bicyclic* graphs. More generally, graphs with cyclomatic number c are called $c-cyclic$ *graphs*.

In the case of trees, since there is no cycle, the degrees of internal vertices decide the number of leaves. For exactly this reason, sometimes for the degree sequences of trees we can only include the internal vertex degrees. Graphical degree sequences that are realized by trees are called *tree degree sequences*. Their characterization is somewhat simpler than the characterization of all graphical degree sequences in Theorem 1.4.1.

Theorem 1.4.2 *A non-increasing sequence* $(d_1, d_2, \ldots, d_n)$ *of positive integers is a tree degree sequence if and only if*

$$\sum_{i=1}^{n} d_i = 2(n-1). \tag{1.2}$$

Proof:

It is clear from (1.1) and the fact that a tree with n vertices has precisely $n - 1$ edges that the condition is necessary. We show by induction that it is also sufficient: this is clear for $n = 2$, where $(1, 1)$ is the only possible sequence that satisfies the condition.

For the induction step, note first that (1.2) can only hold if $d_n = 1$ (since otherwise, the sum is at least equal to $2n$). By the induction hypothesis, there is a tree whose degree sequence is $(d_1 - 1, d_2, d_3, \ldots, d_{n-1})$ (possibly rearranged, if $d_1 = d_2$). Take a vertex in this tree whose degree is $d_1 - 1$, and attach a leaf to it by an edge. The new tree has degree sequence $(d_1, d_2, \ldots, d_n)$. This completes the induction and thus the proof. □

In the study of chemical graph theory, very often one has to consider different degree sequences. In the following, we introduce a way of comparing two degree sequences (on the same number of vertices).

Definition 1.4.1 *Given non-increasing sequences* $\pi' = (d'_1, \ldots, d'_n)$ *and* $\pi'' = (d''_1, \ldots, d''_n)$, π'' *is said to majorize* π', *denoted* $\pi' \lhd \pi''$, *if for* $k \in \{1, 2, \ldots, n - 1\}$

$$\sum_{i=0}^{k} d'_i \leq \sum_{i=0}^{k} d''_i \qquad and \qquad \sum_{i=0}^{n} d'_i = \sum_{i=0}^{n} d''_i.$$

The advantage of defining "majorization" between degree sequences becomes much clearer later on. For now, we just mention that the next lemma has been one of the most frequently used tools in the study of graphs (trees) of different degree sequences.

Lemma 1.4.1 ([117]) *Let $\pi' = (d'_1, \ldots d'_n)$ and $\pi'' = (d''_1, \ldots, d''_n)$ be two non-increasing tree degree sequences. If $\pi' \lhd \pi''$, then there exists a series of (non-increasing) tree degree sequences $\pi^{(i)} = (d_1^{(i)}, \ldots, d_n^{(i)})$ for $1 \leq i \leq m$ such that*

$$\pi' = \pi^{(1)} \lhd \pi^{(2)} \lhd \cdots \lhd \pi^{(m-1)} \lhd \pi^{(m)} = \pi'',$$

and in addition, $\pi^{(i)}$ and $\pi^{(i+1)}$ differ at exactly two entries for every i, say the j-th and k-th entries, $j < k$, where $d_j^{(i+1)} = d_j^{(i)} + 1$ and $d_k^{(i+1)} = d_k^{(i)} - 1$.

Remark 1.4.1 *Lemma 1.4.1 is a more refined version of the original statement in [117]. In this process, each entry stays positive and the degree sequences remain non-increasing. Thereby, each obtained sequence is a tree degree sequence that is non-increasing without rearrangement.*

Sometimes it is advantageous to consider the degrees of the vertices at a given *level*, which is often the set of vertices of the same height in a rooted tree.

Definition 1.4.2 ([94]) *In a rooted tree, the list of multisets L_i of degrees of vertices at height i, starting with L_0 containing the degree of the root vertex, is called the level-degree sequence of the rooted tree.*

1.5 Distance in graphs

There are many interesting concepts related to distances between vertices. We only present the ones most related to our topics.

First, the so-called *distance function* $d_G(v)$ (or simply $d(v)$) of a vertex $v \in V(G)$ is defined as

$$d_G(v) = \sum_{u \in V(G)} d(v, u),$$

the sum of the distances between v and all other vertices. The *centroid* of a graph G is the set of vertices minimizing $d(\cdot)$.

Instead of the sum, if the maximum distance from v is taken, we have the *eccentricity*

$$\mathrm{ecc}(v) = \max_{u \in V(G)} d(u, v).$$

The *radius* of G, $\mathrm{rad}(G)$, is the minimum eccentricity, while the *diameter*, $\mathrm{diam}(G)$, is the maximum. The *center* is the collection of vertices whose eccentricity is exactly $\mathrm{rad}(G)$.

To introduce a concept analogous to the degree sequence, but related to distances, we first define *segments* of a tree (or potentially general graph). A segment of a tree T is a path in T with the property that each of the

FIGURE 1.9
An example graph for independent sets and matchings.

ends is either a leaf or a branching vertex and that all internal vertices of the path have degree 2. The *segment sequence* of T is the non-increasing sequence of the lengths of all segments of T, in analogy to the degree sequence. For example, the quasi-caterpillar in Figure 1.8 has segment sequence $(5, 4, 3, 3, 3, 3, 2, 2, 1, 1, 1, 1, 1, 1, 1, 1)$.

1.6   Independent sets and matchings

An *independent set* in a simple graph G is a subset of $V(G)$ in which no two vertices are adjacent to each other. The *independence number* of G, usually denoted by $\alpha(G)$, is the size of a maximum independent set of G. Similarly, a *matching* in a simple graph G is a set of edges without common vertices. A *maximum matching* is a matching that contains the largest possible number of edges. The *matching number* of G, denoted by $\beta(G)$, is the size of a maximum matching of G. In Chapter 4 we shall see chemical indices based on these concepts. In addition, the extremal graphs with a given matching number has been studied for various concepts related to chemical graph theory. To give a concrete example, consider the graph in Figure 1.9. Its independence number and its matching number are both easily seen to be equal to 2.

Let a component of a graph be called *odd* (*even*) if it has odd (even) number of vertices, and denote the number of odd components of a graph G by $o(G)$. The following result, known as the Tutte-Berge formula (see, e.g., [75]) is a crucial lemma used in the study of the matching number.

Lemma 1.6.1 *Let G be a connected graph of order n. Then*

$$n - 2\beta(G) = \max\{o(G - X) - |X| : X \subseteq V(G)\}.$$

1.7 Topological indices

A large portion of chemical graph theory is concerned with numerical quantities associated with graphs; a famous example is the Wiener index, the sum of the lengths of the shortest paths between all pairs of vertices in the chemical graph representing the non-hydrogen atoms in the molecule. Its history goes back to the papers of Wiener in 1947 [118,119], in which he noted that the boiling temperatures of alkanes can be predicted well by means of a formula that involves the Wiener index (called path number by Wiener). Distance-based invariants such as the Wiener index will be discussed in Chapter 2.

An important aim of chemical graph theory is to come up with useful graph invariants, which have predictive power for chemical properties of the molecule, if computed for the molecular graph. In chemical graph theory, such invariants are called (topological) indices, as the expectation is that the shape of the molecule is the ultimate source of information. The discriminating power of an index is high if different graph shapes tend to result in distinct index values. The range of an index limits its discriminating power. We do not even try to form a complete list of topological indices as the number of such indices is quite large. Instead, we try to present useful techniques in the study of chemical graph theory through some representative examples.

As mentioned earlier, the following chapter will be devoted to distance-based indices, in particular the Wiener index and its variants. Randić [90] introduced another very influential index, which is now named after him. Previously, it was called the branching index or connectivity index. The Randić index, which is the prototype of degree-based indices, and its variants will be reviewed in Chapter 3. The Merrifield-Simmons index of a graph is the number of its independent vertex sets [77], and the Hosoya index (also called topological Z index) of a graph is the number of its matchings [49]. We discuss these topics in Chapter 4. Finally, Chapter 5 will be devoted to graph spectra and invariants based on spectra, specifically the graph energy that was introduced by Gutman [33].

Many topological indices satisfy a natural monotonicity property. In the following, we will write $G - A$ for the graph resulting from G by removing the vertices or edges contained in the set A. In particular, we will simply write $G - v$ and $G - e$ for the graph obtained from G by removing vertex v (edge e, respectively). We also write $G + e$ for the graph that is obtained from G by adding an edge e.

A graph invariant F that assigns a value $F(G)$ to every graph G is said to be increasing if

$$F(G) > F(G - e)$$

for every edge e of a (non-edgeless) graph G, or equivalently

$$F(G + e) > F(G)$$

for every edge e added to a non-complete graph G. Likewise, F is called decreasing if

$$F(G) < F(G - e)$$

for every edge e of a (non-edgeless) graph G, or equivalently

$$F(G + e) < F(G)$$

for every edge e added to a non-complete graph G. In all cases, we assume that $G, G - e$ and $G + e$ are connected if F is only defined for connected graphs.

The following simple, but important, observation will be used frequently.

Proposition 1.7.1 • *If a graph invariant F is increasing, then*

$$F(E_n) < F(G) \leq F(K_n)$$

for all graphs G with n vertices. If F is decreasing, then

$$F(E_n) \geq F(G) \geq F(K_n)$$

for all graphs G with n vertices. In both cases, equality only holds if G is edgeless (complete, respectively). If F is only defined for connected graphs, then only the second inequality holds in each case.

• *Suppose that F is increasing. For every positive integer n, there exists an n-vertex tree T such that*

$$F(T) \leq F(G)$$

for all connected graphs G with n vertices. Likewise, if F is decreasing, then for every positive integer n, there exists an n-vertex tree T such that

$$F(T) \geq F(G)$$

for all connected graphs G with n vertices.

We conclude this introductory chapter with a general lemma, taken from [15], that provides information on graphs with a given matching number that maximize or minimize a graph invariant under the conditions of Proposition 1.7.1. It provides a first flavor of the type of results to be found in this book.

For two graphs G and H, $G \cup H$ denotes the vertex-disjoint union of G and H. $G + H$ (called the join of G and H) denotes the graph obtained from $G \cup H$ by adding all edges between every vertex of G and every vertex of H.

Lemma 1.7.1 *If the graph invariant F is increasing (decreasing), the minimum (maximum) of $F(G)$ among all connected graphs of order n and matching number β is achieved by a graph of the form*

$$\widehat{G} = K_s + (K_{n_1} \cup K_{n_2} \cup \cdots \cup K_{n_t})$$

for some s and t with $s + n_1 + \ldots + n_t = n$.

Proof:

We only consider the case that F is decreasing. Let $\widehat{G}$ be a connected graph having minimum $F(G)$ among all connected graphs with n vertices and matching number β. By Lemma 1.6.1, there exists a set $\widehat{X} \subseteq V(\widehat{G})$ such that

$$n - 2\beta = \max\{o(\widehat{G} - X) - |X| : X \subseteq V(\widehat{G})\} = o(\widehat{G} - \widehat{X}) - |\widehat{X}|.$$

For simplicity, let $|\widehat{X}| = s$ and $o(\widehat{G} - \widehat{X}) = t$, then $n - 2\beta = t - s$.

First suppose that $s = 0$, then

$$t = n - 2\beta = o(\widehat{G} - \widehat{X}) = o(\widehat{G}) \leq 1.$$

When $t = 0$ or 1, $\beta = \lfloor \frac{n}{2} \rfloor$ in this case and the minimum $F(G)$ is achieved by the complete graph.

Now let $s \geq 1$ and hence $t \geq 1$. We claim that there is no even component in $\widehat{G} - \widehat{X}$. Otherwise, by adding an edge in $\widehat{G}$ between a vertex of an even component and a vertex of an odd component of $\widehat{G} - \widehat{X}$, we obtain $\widehat{G}'$ with

$$n - 2\beta(\widehat{G}') \geq o(\widehat{G}' - \widehat{X}) - |\widehat{X}| = o(\widehat{G} - \widehat{X}) - |\widehat{X}| = n - 2\beta \geq n - 2\beta(\widehat{G}').$$

Thus $\beta(\widehat{G}') = \beta$ and by the condition on $F(\cdot)$, we have $F(\widehat{G}') < F(\widehat{G})$, a contradiction.

Now $\widehat{G} - \widehat{X}$ contains only odd components, denoted by $G_1, \ldots, G_t$. Since the addition of edges will decrease $F(G)$, we can assume that $G_1, G_2, \ldots, G_t$, and the subgraph induced by $\widehat{X}$ are all complete and each vertex of $G_1, G_2, \ldots, G_t$ is adjacent to every vertex in $\widehat{X}$. Let $n_i = |V(G_i)|$ for $i = 1, 2, \ldots, t$, the conclusion follows. $\qquad\square$

Later, we make use of Lemma 1.7.1 to characterize or partially characterize extremal graphs of interest in chemical graph theory.

Exercises

1. Determine whether the following sequences are graphical: $(5, 3, 2, 2, 2, 1)$, $(5, 3, 3, 2, 2, 1)$, $(5, 5, 5, 3, 2, 2)$.

2. Prove Lemma 1.4.1.

3. Prove: a non-increasing sequence $(d_1, d_2, \ldots, d_n)$ of positive integers is the degree sequence of a unicyclic graph if and only if

$$\sum_{i=1}^{n} d_i = 2n$$

and the sequence contains at least three elements greater then or equal to 2.

4. Prove: for every tree T with n vertices, the independence number $\alpha(T)$ and the matching number $\beta(T)$ satisfy

$$\alpha(T) + \beta(T) = n.$$

5. Prove Proposition 1.7.1.

6. Prove: if f is an increasing function on the non-negative integers, then

$$\sum_{v \in V(G)} f(\deg(v))$$

is an increasing invariant, and

$$\sum_{v \in V(G)} f(d(v))$$

is a decreasing invariant.

7. The eccentric connectivity index of a connected graph G is defined by

$$EC(G) = \sum_{v \in V(G)} \deg(v) \, \mathrm{ecc}_G(v).$$

Is this index an increasing or a decreasing invariant?

2

Distance in graphs and the Wiener index

2.1 An overview

The *Wiener index* is commonly considered as the most classic and widely used distance-based index in chemical graph theory. It is named after the chemist Harry Wiener, who proposed this concept (originally calling it the path number) in 1947 [118, 119]. Given a graph G, the Wiener index of G is defined as

$$W(G) = \sum_{\{u,v\} \subseteq V(G)} d(u,v),$$

where $d(u,v)$ is the distance between vertices u and v in G.

For example, in Figure 2.1, we have a graph on 6 vertices, the $\binom{6}{2} = 15$ distances between pairs of vertices are

$$1, 1, 1, 1, 1, 1, 1, 2, 2, 2, 2, 2, 2, 2, 2.$$

Hence, the Wiener index is $1 \times 7 + 2 \times 8 = 23$.

FIGURE 2.1
A graph with Wiener index 23.

It was noted by Wiener that the Wiener index (of a molecular graph), combined with another quantity now known as the Wiener polarity index, can be used in a formula to predict the boiling points of alkane molecules quite accurately. The mathematical examination of the Wiener index, as the sum of distances between vertices, probably started from [25] or earlier without knowledge of the applications in chemistry.

Of all graphs, trees received particular attention in the study of the Wiener index because of the many acyclic molecular structures in applications. An early informative survey is [24]. The Wiener index of general graphs and graphs with various given parameters have also been studied.

Among general trees of given order, the extremal trees that maximize or

minimize the Wiener index have been characterized. The following theorem
will be proved later in this chapter:

Theorem 2.1.1 ([25], [73] Ex. 6.23) *Among all trees of the same order,
the star minimizes the Wiener index and the path maximizes the Wiener index.*

The star and the path are also extremal trees for many other topological
indices, as we will see later. Since the degree of a vertex in the molecular graph
correspond to *valence* of an atom in a molecule, it is natural to examine the
behavior of the Wiener index in graphs under various vertex degree restrictions. In particular, Fischermann, Hoffmann, Rautenbach, Székely and Volkmann [30], and independently Jelen and Triesch [59] characterized the trees
with minimum Wiener index among all trees (of given order) with a bounded
maximum degree, and the trees with maximum Wiener index among all trees
(of given order) whose vertex degrees are 1 or k.

The study in [30] was generalized to trees with a given degree sequence
in [111] and [128], respectively, where it is shown that the minimum Wiener
index is attained by the greedy tree, which is—intuitively speaking—the most
compact tree with a given degree sequence. Let us formally define greedy trees.

Definition 2.1.1 (Greedy trees) *With given vertex degrees, the greedy tree
is constructed through the following "greedy algorithm":*

i) Label the vertex with the largest degree as v (the root);

*ii) Label the neighbors of v as v_1, v_2, ..., assign the largest degrees available
to them such that $\deg(v_1) \geq \deg(v_2) \geq \cdots$;*

*iii) Label the neighbors of v_1 (except v) as v_{11}, v_{12}, ... such that they take
all the largest degrees available and that $\deg(v_{11}) \geq \deg(v_{12}) \geq \cdots$, then do
the same for v_2, v_3, ...;*

*iv) Repeat (iii) for all the newly labeled vertices, always start with the
neighbors of the labeled vertex with largest degree whose neighbors are not
labeled yet.*

For example, Figure 2.2 shows a greedy tree with degree sequence

$$\{4, 4, 4, 4, 3, 3, 3, 3, 3, 2, 1, \ldots, 1\}.$$

For technical reasons, the following *level-greedy tree* is also defined when
considering trees with a given level-degree sequence.

Definition 2.1.2 (Level-greedy trees) *For $i = 0, 1, \ldots, H$, let multisets
$\{a_{i1}, a_{i2}, \ldots, a_{i\ell_i}\}$ of non-negative numbers be given such that $\ell_0 = 1$ and*

$$\ell_{i+1} = \sum_{j=1}^{\ell_i} a_{ij}.$$

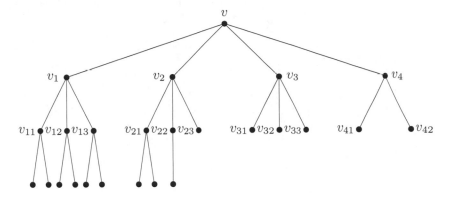

FIGURE 2.2
A greedy tree.

Assume that the elements of each multiset are sorted, i.e., $a_{i1} \geq a_{i2} \geq \cdots \geq a_{i\ell_i}$. The level-greedy tree corresponding to this sequence of multisets is the rooted tree whose j-th vertex at level i has outdegree a_{ij}.

Likewise, if sorted multisets $\{a_{i1}, a_{i2}, \ldots, a_{i\ell_i}\}$ of non-negative numbers are given for $i = 0, 1, \ldots, H$ such that $\ell_0 = 2$ and

$$\ell_{i+1} = \sum_{j=1}^{\ell_i} a_{ij},$$

then the level-greedy tree corresponding to this sequence of multisets is the edge-rooted tree (i.e., there are two vertices at level 0, connected by an edge) whose j-th vertex at level i has outdegree a_{ij}.

Every greedy tree is clearly also level-greedy (with respect to any root vertex), but the converse is not true (i.e., a tree can be level-greedy with respect to a certain root without being a greedy tree). For example, Figure 2.3 shows a level-greedy tree corresponding to the following sequence of multisets: $\{a_{01} = 3\}$, $\{a_{11} = 4, a_{12} = 2, a_{13} = 1\}$, $\{2, 2, 1, 1, 1, 0, 0\}$, $\{1, 1, 1, 0, 0, 0, 0\}$ and $\{0, 0, 0\}$.

The following theorem, which will be proved later, holds:

Theorem 2.1.2 ([111, 128]) *Among all trees with a given degree sequence (and hence given order), the greedy tree minimizes the Wiener index.*

To find the maximum Wiener index among trees with a given degree sequence is a much more difficult question. It has been shown in as early as 1993 that the extremal tree has to be a caterpillar [96]. Further studies can be found in [97] and [127], where it was noted that the specific characteristics of such extremal trees depend on the particular degree sequence. This question was also examined as a quadratic assignment problem in [14], where an efficient algorithm was provided.

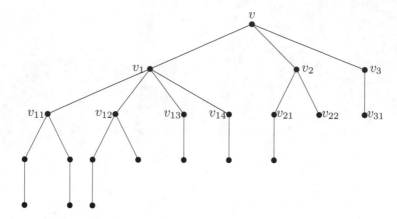

FIGURE 2.3
A level-greedy tree.

The extremality of greedy trees can be further extended to more general distance-based indices defined through a function on the distances:

Theorem 2.1.3 ([94]) *Let $f(x)$ be any non-negative, non-decreasing function of x. Then the graph invariant*

$$W_f(T) = \sum_{\{u,v\} \subseteq V(T)} f(d(u,v))$$

is minimized by the greedy tree among all trees with given degree sequence.
    Likewise, if $f(x)$ is any non-negative, non-increasing function of x, then the graph invariant W_f is maximized by the greedy tree among all trees with given degree sequence.

With different choices of f the above theorem leads to the extremality of the greedy tree with respect to many variations of the Wiener index. The following are some of the most well-known ones:

- In 1993, the *hyper-Wiener index* [91] was introduced for trees and later generalized to cyclic graphs [64]:

$$WW(T) = \frac{1}{2} \sum_{\{u,v\} \subseteq V(T)} \left(d(u,v) + d(u,v)^2 \right),$$

- the *Harary index* was defined in [58, 88]:

$$H(T) = \sum_{\{u,v\} \subseteq V(T)} \frac{1}{d(u,v)},$$

- and the *terminal Wiener index* was proposed in [36]:

$$TW(T) = \sum_{\{u,v\} \subseteq \ell(T)} d(u,v),$$

where $\ell(T)$ stands for the set of leaves of T. In addition to its application in chemistry, the terminal Wiener index, being simply the sum of distances between leaves, is also found to be of importance in the study of phylogenetic trees and is known there as the *gamma index* [101].

Other variations of distances between vertices that have been studied include the sum of distances between internal vertices and leaves [113], the sum of the *eccentricities* and equivalently the *average eccentricity*. The star, path, greedy trees, and caterpillars continue to be extremal with respect to these indices. Some lesser known distance-based indices are mentioned in [120].

The notion of majorization between degree sequences provides a means of comparing the extremal trees (in this case, the greedy trees) for different degree sequences:

Theorem 2.1.4 ([108]) *Let $f(x)$ be any non-negative, non-decreasing function of x, and let π and π' be two degree sequences of trees of the same length such that π' majorizes π. If $G(\pi)$ and $G(\pi')$ are the greedy trees associated with π and π', respectively, we have*

$$W_f(G(\pi)) \geq W_f(G(\pi')).$$

Likewise, if f is a non-negative, non-increasing function, then

$$W_f(G(\pi)) \leq W_f(G(\pi')).$$

From Theorem 2.1.4, many interesting extremal results on distance-based indices follow as immediate corollaries, as we will see later in this chapter.

2.2 Properties related to distances

One of the first questions one may have for the Wiener index is probably how to compute it for a given graph. For a general graph, one may simply sum up the distances from each vertex to the rest of the vertices, through a "breadth-first search" algorithm. For a graph with n vertices and m edges, the time complexity is $O(nm)$ for this approach. For trees, the Wiener index can be computed through a linear-time algorithm, based on the following simple but useful observation that already appears in Wiener's original paper [119].

Proposition 2.2.1 *In a tree T, the Wiener index can be computed as*

$$W(T) = \sum_{uv \in E(T)} n_{uv}(v) \cdot n_{uv}(u)$$

where $n_{uv}(v)$ ($n_{uv}(u)$) is the number of vertices closer to v (u) than to u (v) in T.

Proof:

Note that the Wiener index is the sum of distances between vertices, which is the number of edges between pairs of vertices. In other words, to compute the Wiener index, one counts the number of times each edge uv is used as part of a path between a pair of vertices, and then takes the sum of these numbers over all edges.

On the other hand, the number of paths containing a particular edge uv is the number of ways to choose two end vertices of this path, one on each side of the edge. The number of choices here is exactly the product of the numbers of vertices on two sides of this edge. □

The very same idea applies to the computation of similar distance-based indices. For this purpose, we let $n'_{uv}(v)$ and $n'_{uv}(u)$ ($n''_{uv}(v)$ and $n''_{uv}(u)$) denote the number of internal vertices (leaves) closer to v than to u in a given tree T. The following observations follow from exactly the same argument. We leave the proofs to interested readers as exercises.

Proposition 2.2.2 *In a tree T, the sum of distances between internal vertices is*

$$\sum_{uv \in E(T)} n'_{uv}(v) \cdot n'_{uv}(u).$$

Proposition 2.2.3 *In a tree T, the sum of distances between leaves is*

$$\sum_{uv \in E(T)} n''_{uv}(v) \cdot n''_{uv}(u).$$

Very often in the study of extremal structures that maximize or minimize a topological index, we examine the impact of some graph transformation on the value of that particular index. In the case of the Wiener index, we first note the following.

Lemma 2.2.1 *Let R be a graph with two vertices u and w such that*

$$d_R(u) > d_R(w),$$

and let S be a graph with at least two vertices and a specified vertex (root) v. Consider the graphs G obtained by attaching S to w (identifying v with w) and G' obtained by attaching S to u (identifying v with u). Then

$$W(G') > W(G).$$

Proof:

Consider the distances between pairs of vertices in G and G', respectively. Through the operation that transforms G to G' the distances between any two vertices both in R or both in $S - v$ stay the same. So we only have to consider the distances between pairs of vertices such that one is in $S - v$ and the other is in R.

In G, the sum of these distances is

$$d_S(v) \cdot |V(R)| + d_R(w) \cdot (|V(S)| - 1)$$

as the distance from any vertex in S to v is counted as many times as the number of paths from w to any vertex in R and similarly for the distance from any vertex in R to w.

For exactly the same reason, the sum of these distances in G' is

$$d_S(v) \cdot |V(R)| + d_R(u) \cdot (|V(S)| - 1).$$

Thus,
$$W(G') - W(G) = (d_R(u) - d_R(w)) \cdot (|V(S)| - 1) > 0,$$

as claimed. □

Intuitively, Lemma 2.2.1 states that attaching the same "rooted graph" (S) to a vertex that is farther away from the rest of the graph (R) results in greater value for the Wiener index. Following the same logic, one can show the following generalized version of Lemma 2.2.1.

Lemma 2.2.2 *Let R be a graph with two vertices u and w such that*

$$d_R(u) > d_R(w),$$

and let S_i be a graph with at least two vertices and a specified vertex (root) v_i (for $i = 1, 2$) with
$$|V(S_1)| > |V(S_2)|.$$

Consider the graphs G obtained by attaching S_1 to w, S_2 to u and G' obtained by attaching S_1 to u, S_2 to w (Figure 2.4). Then

$$W(G') > W(G).$$

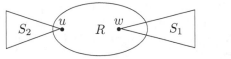 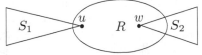

FIGURE 2.4
The graphs G (left) and G' (right).

Proof:

Following exactly the same arguments, the only distances changed from G to G' are those between pairs of vertices in $S_i - v_i$ and R, respectively, for $i = 1, 2$.

The sum of these distances is

$$d_{S_1}(v_1) \cdot |V(R)| + d_R(w) \cdot (|V(S_1)| - 1) + d_{S_2}(v_2) \cdot |V(R)| + d_R(u) \cdot (|V(S_2)| - 1)$$

in G and

$$d_{S_1}(v_1) \cdot |V(R)| + d_R(u) \cdot (|V(S_1)| - 1) + d_{S_2}(v_2) \cdot |V(R)| + d_R(w) \cdot (|V(S_2)| - 1)$$

in G'. Hence

$$W(G') - W(G) = (d_R(u) - d_R(w)) \cdot (|V(S_1)| - |V(S_2)|) > 0$$

as claimed.   □

Again, intuitively Lemma 2.2.2 simply says that attaching more vertices to the further end of a graph results in greater value for the Wiener index. It is interesting that the statement holds regardless of the specific structures of S_1 and S_2.

In addition to graph transformations, it is also important to note that the Wiener index is a "monotone" invariant in the following sense.

Lemma 2.2.3 *For a connected graph on n vertices and $n - 1 \le m < \binom{n}{2}$ edges, adding an edge uv between two non-adjacent vertices u and v will decrease the Wiener index.*

Likewise, removing an edge (while keeping the graph connected) will increase the Wiener index.

Proof:

Let G be a connected graph on n vertices and m edges, with $m < \binom{n}{2}$. Suppose u and v are vertices of G that are not adjacent. Thus the distance between u and v in G is at least 2. Consider now the graph $G' = G + \{uv\}$. The distance between u and v in G' is 1. And it is easy to see that the distance between any other pair of vertices can only decrease. Thus $W(G') < W(G)$ as claimed.

Similarly, removing an edge uv from G will increase the distance between u and v from 1 to at least 2, while the distance between any other pair of vertices can only increase.   □

While the Wiener index deals with the sum of distances, either between all pairs of vertices or from all vertices to a single vertex, the eccentricity deals with largest distance from a vertex. The following fact helps identify this largest distance from a diametral path in a given tree. We write $P(u, v)$ for the unique path between vertices u and v in a tree.

Lemma 2.2.4 *For a tree T and a longest path $P(u, v)$ in T, the eccentricity of any vertex w is obtained by at least one of $P(u, w)$ and $P(v, w)$.*

Proof:

Suppose, for contradiction, that the eccentricity of w is obtained by a path that is longer than both $P(u, w)$ and $P(v, w)$. Consider two cases:

- If w is on $P(u, v)$ and the eccentricity is obtained by a path $P(w, w')$ that is longer than both $P(u, w)$ and $P(v, w)$: Assume, without loss of generality, that
$$V(P(w, w')) \cap V(P(w, v)) = \{w\}.$$
Then, $P(w', v)$ is a longer path than $P(u, v)$, a contradiction.

- If w is not on $P(u, v)$ and the eccentricity of w is obtained by a path $P(w, w'')$ that is longer than both $P(u, w)$ and $P(v, w)$: Assume, without loss of generality, that
$$V(P(w, w'')) \cap V(P(w, v)) = \{w\}.$$
Then, $P(w'', v)$ is a longer path than $P(u, v)$, a contradiction. Note that this argument holds regardless of whether $P(w, w'')$ has any overlap with $P(u, v)$.

$\square$

2.3 Extremal problems in general graphs and trees

We now consider one of the most thoroughly studied questions in chemical graph theory, the extremal problem, for the Wiener index and some closely related distance-based indices. We will start with general graphs and trees. More specific classes of structures will be considered later.

2.3.1 The Wiener index

First of all, it is a trivial exercise to see that the Wiener index is minimized by the complete graphs and stars.

Proposition 2.3.1 *Among connected graphs on n vertices, the Wiener index is minimized by the complete graph (with Wiener index $\binom{n}{2}$).*

Proof:

For a connected graph on n vertices, the distance between any of the $\binom{n}{2}$ pairs of vertices is at least 1, obtained by and only by the complete graph. This is also a direct consequence of Lemma 2.2.3. $\square$

Proposition 2.3.2 *Among trees on n vertices, the Wiener index is minimized by the star with Wiener index $2\binom{n-1}{2} + (n-1) = (n-1)^2$.*

Proof:

For a tree on n vertices, there are exactly $n - 1$ edges or pairs of vertices at distance 1. For the remaining $\binom{n-1}{2}$ pairs of vertices, they are at distance at least 2 from each other. This is obtained by and only by a star. □

On the other hand, the path is known to maximize the Wiener index among trees of given order. This can be established by arguing that the "largest" collection of distances between vertices consists of at least $(n - 1)$ 1s, $(n - 2)$ 2s, etc., and that this can only be achieved by a path. In the following we show, instead of such a "numerical" proof, an approach that utilizes the structural properties we established earlier.

Proposition 2.3.3 *Among trees on n vertices, the Wiener index is maximized by the path.*

Proof:

Suppose, for contradiction, that T is a tree on n vertices that maximizes the Wiener index and that T is not a path.

Let $P(u, v)$ be a longest path in T between leaves u and v, and let its vertices be denoted $u = v_0, v_1, v_2, \ldots, v_k, v_{k+1} = v$. Since T is not a path, there exists a vertex on $P(u, v)$, say v_i, with the smallest subscript i such that v_i is of degree at least 3. It is obvious that i is at least 1 and at most k.

Let S be the component in $T - v_i v_{i+1} - v_i v_{i-1}$ that contains v_i (Figure 2.5). Note that S contains at least two vertices.

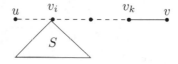

FIGURE 2.5
The path $P(u, v)$, v_i, and S.

Let $R = T - (S - v_i)$. It is a simple exercise to verify that $d_R(u) > d_R(v_i)$. In fact, the sum of distances between u and the vertices on $P(u, v_i)$ is exactly the same as the sum of distances between v_i and these vertices. The distance between u and any other vertex in R is exactly the distance between v_i and that vertex plus i.

We now consider the tree T' obtained from T by detaching S from v_i and reattaching it to u. A direct application of Lemma 2.2.1 shows that $W(T') > W(T)$, a contradiction. □

As an immediate corollary of Lemma 2.2.3, the maximum Wiener index of a connected graph on n vertices must be achieved by a tree (i.e., a connected graph with the least number of edges). Then, we have the following from Proposition 2.3.3 above.

Corollary 2.3.1 *Among connected graphs on n vertices, the Wiener index is maximized by the path.*

2.3.2 The distances between leaves

Recall that the sum of all distances between leaves is

$$\Gamma(T) = \sum_{v,u\in\ell(T)} d(v,u)$$

where $\ell(T)$ denotes the set of leaves of T. This was proposed independently as the *Gamma index* [101] (from the study of "tree bisection and reconnection neighborhood" in phylogeny reconstruction) and the *terminal Wiener index* [36] (for similar purpose as the original Wiener index).

It is not difficult to find examples of trees, as shown in Figure 2.6, where the Wiener index and the terminal Wiener index do not share any monotonic functional relation. Here T_1 and T_2 are two different trees, not only of the same order but also with the same degree sequence, and it is not hard to verify that

$$W(T_1) > W(T_2)$$

and

$$\Gamma(T_1) < \Gamma(T_2).$$

FIGURE 2.6
The trees T_1 on the left and T_2 on the right.

On the other hand, for two k-ary trees T and T' (trees whose internal vertices all have degree k), the following is known [112]. We leave the proof to interested readers.

$$W(T) - W(T') = \left(\frac{k-1}{k-2}\right)^2 (\Gamma(T) - \Gamma(T')). \tag{2.1}$$

Such a correlation (2.1) implies that, for k-ary trees, the extremal structures that maximize or minimize the terminal Wiener index coincide with those that maximize or minimize the Wiener index. Such extremal structures will be discussed in more detail in later sections of this chapter. For the moment we examine more basic extremal structures with respect to the sum of distances between leaves.

As far as extremal problems with respect to $\Gamma(\cdot)$ are concerned, it seems reasonable to restrict our attention to trees with a fixed number of leaves as only distances between leaves are considered. First, we claim the following for trees that minimize the terminal Wiener index. Recall that a starlike tree is a tree with only one branching vertex (vertex of degree at least 3). See Figure 2.7.

FIGURE 2.7
A starlike tree.

Proposition 2.3.4 *Among trees on n vertices and $\ell \geq 3$ leaves, every starlike tree attains the minimum terminal Wiener index $(n-1)(\ell-1)$.*

Proof:

By Proposition 2.2.3, the terminal Wiener index is represented by

$$\sum_{uv \in E(T)} n''_{uv}(v) \cdot n''_{uv}(u).$$

Since there are ℓ leaves, the product

$$n''_{uv}(v) \cdot n''_{uv}(u) = x \cdot (\ell - x)$$

for some $x \geq 1$ for any edge uv. Hence it is at least $1 \cdot (\ell - 1)$ for each of the $(n-1)$ edges. Consequently,

$$\Gamma(T) \geq (n-1) \cdot (\ell - 1)$$

with equality if and only if the tree has exactly one branching vertex (for $\ell \geq 3$) or no branching vertex (in which case $\ell = 2$ and T is a path). $\quad\square$

Next, we examine the maximum terminal Wiener index in trees with given number of vertices and leaves. Intuitively, such extremal trees would place the leaves as far apart from each other as possible. For this purpose we introduce the *r-dumbbell* as a tree obtained by attaching pendant edges to two ends of a path of length r, such that the numbers of pendant edges at the two ends differ by at most 1. See Figure 2.8.

Proposition 2.3.5 *Among trees on n vertices with ℓ leaves, the $(n-\ell-1)$-dumbbell attains the maximum terminal Wiener index*

$$\ell(\ell-1) + (n - \ell - 1) \left\lfloor \frac{n+1}{2} \right\rfloor \left\lceil \frac{n-1}{2} \right\rceil.$$

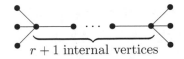

FIGURE 2.8
An r-dumbbell.

Proof:

Similar to the last proof, consider the formula

$$\Gamma(T) = \sum_{uv \in E(T)} n''_{uv}(v) \cdot n''_{uv}(u)$$

from Proposition 2.2.3.

Since there are ℓ leaves, there are ℓ pendant edges. For each of these pendant edges the product $n''_{uv}(v) \cdot n''_{uv}(u)$ contributes $1 \cdot (\ell - 1)$ to $\Gamma(T)$.

For any of the $n - \ell - 1$ remaining edges, the product $n''_{uv}(v) \cdot n''_{uv}(u)$ is at most $\lfloor \frac{n+1}{2} \rfloor \lceil \frac{n-1}{2} \rceil$. Thus,

$$\Gamma(T) \leq \ell(\ell - 1) + (n - \ell - 1) \left\lfloor \frac{n+1}{2} \right\rfloor \left\lceil \frac{n-1}{2} \right\rceil$$

with equality if and only if the tree is an $(n - \ell - 1)$-dumbbell. $\qquad\square$

2.3.3   Distance between internal vertices

Probably inspired by the distance between all vertices and the distance between all leaves, a concept named the *spinal index* $S(\cdot)$ was defined as the sum of the distances between all internal vertices [6]. It is easy to see that the spinal index of a tree T is essentially the Wiener index of the "skeleton" $T - \ell(T)$ of T, hence it is not surprising that many studies of the Wiener index can be easily generalized to the spinal index [17]. We briefly mention two simple observations here.

Any tree T that is not a star has at least two internal vertices and hence $S(T) > 0$. Since $S(T) = 0$ for a star, the following is trivial.

Proposition 2.3.6 *Among trees with given order, the star is the unique tree that minimizes $S(T)$.*

Similarly, a tree T has at most $|V(T)| - 2$ internal vertices (with the upper bound achieved if and only if T is a path) and $S(T) = W(T')$ is maximized when T' is a path.

Proposition 2.3.7 *Among trees with given order, the path is the unique tree that maximizes $S(T)$.*

2.3.4   Distance between internal vertices and leaves

Of course, the Wiener index is simply the sum of the Gamma index, the spinal index, and the sum of all distances between internal vertices and leaves in a tree. This last concept is denoted by

$$K(T) = \sum_{u \in V(T) - \ell(T), v \in \ell(T)} d(u, v)$$

for a tree T.

It is easy to show that the star stays as an extremal structure for $K(T)$.

Proposition 2.3.8 *For any tree T on n vertices we have*

$$K(T) \geq n - 1$$

with equality if and only if T is a star.

Proof:

For a tree T, $K(T)$ is the sum of the distances between $n - |\ell(T)|$ internal vertices and $|\ell(T)|$ leaves, hence a total of

$$(n - |\ell(T)|)|\ell(T)|$$

distances.

Note that this product is at least $1 \cdot (n - 1)$ (only in the case of a star, where there is only one internal vertex) and each distance is at least 1, hence

$$K(T) \geq 1 \cdot (n - 1) \cdot 1$$

with equality if and only if T is a star. $\square$

On the other hand, the path is not extremal with respect to $K(\cdot)$. To characterize the (perhaps a little surprising) structure that maximizes $K(T)$ among trees of given order, first note the following fact.

Lemma 2.3.1 *For any $u \in \ell(T)$ and $k \geq 2$, there exists an internal vertex v with $d(u, v) = k$ only if there exists an internal vertex w such that $d(w, v) = k - 1$.*

Proof:

This can be verified by simply taking the unique neighbor of v on the path $P(v, u)$ to be w. $\square$

Proposition 2.3.9 *Among trees on n vertices, the maximum $K(T)$ is $\sim \frac{2n^3}{27}$, obtained by a tree formed by attaching a total of $n - m$ pendant edges to the* end vertices of a path of length $m - 1$ *(with at least one pendant edge at each end), where*

$$\frac{2}{3}n \sim m = \left\lfloor \frac{(n-1) + \sqrt{(n-1)^2 + 3n}}{3} \right\rfloor \quad or \quad \left\lceil \frac{(n-1) + \sqrt{(n-1)^2 + 3n}}{3} \right\rceil.$$

In particular, the $(m - 1)$-dumbbell is one of these extremal structures (Figure 2.8 with $r = m - 1$).

Proof:

With given number m of internal vertices, Lemma 2.3.1 implies that the sum of distances between one leaf u and all internal vertices is

$$\sum_{v \in V(T) - \ell(T)} d(u, v) \le \sum_{i=1}^{m} i = \frac{1}{2}m(m+1). \qquad (2.2)$$

By applying (2.2) to every leaf of T, we have

$$K(T) \le \frac{1}{2}m(m+1)(n - m) =: f(m)$$

with equality if and only if T is formed by attaching a total of $n - m$ pendant edges to a path of length $m - 1$ (with at least one pendant edge at each end).

Now, in order to find the extremal value of $f(m)$, we set

$$f'(m) = \frac{1}{2}\left(-3m^2 + 2(n-1)m + n\right)$$

to be 0. It then follows from basic calculus that $f(m)$ is maximized when

$$m = \left\lfloor \frac{(n-1) + \sqrt{(n-1)^2 + 3n}}{3} \right\rfloor \quad or \quad \left\lceil \frac{(n-1) + \sqrt{(n-1)^2 + 3n}}{3} \right\rceil.$$

$\square$

It is interesting to note that, unlike most other known results regarding general extremal trees, the extremal structure in this case is evidently not unique.

2.3.5 Sum of eccentricities

Recall that the *eccentricity* of a vertex v in a connected graph G is defined in terms of the distance function as

$$\mathrm{ecc}_G(v) := \max_{u \in V(G)} d(u, v).$$

We are interested in the *total eccentricity* of a tree T, defined as the sum of the vertex eccentricities:

$$\text{Ecc}(T) := \sum_{z \in V(T)} \text{ecc}_T(z).$$

It was shown in as early as 2004 [20] that the path maximizes $\text{Ecc}(T)$ among trees of given order. We provide a brief justification for the extremality of the path and leave the computation of the extremal values as an exercise to interested readers.

Proposition 2.3.10 *Among trees of a given order, the path maximizes* $\text{Ecc}(T)$.

Proof:

Suppose for contradiction that there is a tree T with the maximum $\text{Ecc}(T)$, and that T is not a path. Let $P(u, v)$ be a longest path in T with vertices $u = v_0, v_1, \ldots, v_k, v_{k+1} = v$. Similar to the proof of Proposition 2.3.3, let v_i be the vertex with the smallest i that has a neighbor w not on $P(u, v)$ (Figure 2.9).

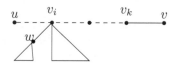

FIGURE 2.9
The path $P(u, v)$, v_i, and w.

Consider now the tree $T' = T - v_i w + uw$. By Lemma 2.2.4 it is not hard to see that the eccentricity of every vertex stayed the same or increased from T to T'. □

It is also not difficult to show that the star minimizes $\text{Ecc}(T)$ among trees of a given order.

Proposition 2.3.11 *For any tree T on $n > 2$ vertices, we have*

$$\text{Ecc}(T) \geq 1 + 2(n-1) = 2n - 1$$

with equality if and only if T is a star.

Proof:

First note that, for any tree with at least 3 vertices, there is at most one vertex with eccentricity 1 (it has to be adjacent to every other vertex in the tree). For every other vertex the eccentricity is at least 2. Thus, we have

$$\text{Ecc}(T) \geq 1 + 2(n-1) = 2n - 1$$

with equality if and only if T is the star. $\square$

2.4 The Wiener index of trees with a given degree sequence

As already mentioned earlier in the overview section, trees with a given degree sequence are an important class of structures in the study of the Wiener index and other distance-based indices. The extremal problems for the Wiener index in trees with a given degree sequence have been extensively studied. The following has been established in [110] and [128].

Theorem 2.4.1 *Among trees with a given degree sequence, the Wiener index is minimized by the greedy tree.*

We skip the proof here, as later in this chapter we will establish a general statement regarding the extremality of greedy trees with respect to distance-based indices including the Wiener index.

On the other hand, the problem of maximizing the Wiener index among trees with a given degree sequence turned out to be a difficult problem. It has been proved several times, probably first by Shi [96], that a tree with maximal Wiener index for a given degree sequence must be a caterpillar. We will formally introduce this statement later as part of a more general result, see Proposition 2.6.1.

To actually characterize the caterpillar, with a given degree sequence, that maximizes the Wiener index, is essentially impossible as the extremal caterpillar depends on the specific degree sequence. To at least partially characterize the extremal caterpillar, we first introduce the *decremented degree sequence*

$$\mathbf{b} = (b_1, b_2, \ldots, b_k) := (d_1 - 1, d_2 - 1, \ldots, d_k - 1)$$

for a given degree sequence

$$\mathbf{d} = \mathbf{d}_T = (d_1, d_2, \ldots, d_k \geq 2, d_{k+1} = 1, \ldots, d_n = 1).$$

Given a caterpillar T, let the non-leaf vertices in the backbone be $v_1, v_2, \ldots, v_k$. The Wiener index of T is given by

$$W(T) = (n-1)^2 + q(\mathbf{x}), \tag{2.3}$$

where $q(\mathbf{x})$ is the quadratic form

$$q(\mathbf{x}) = \frac{1}{2} \sum_{i=1}^{k} \sum_{j=1}^{k} |i - j| x_i x_j = \sum_{1 \leq i < j \leq k} (j - i) x_i x_j, \tag{2.4}$$

with **x** being the column vector

$$\mathbf{x} = (x_1, x_2, \ldots, x_k)^t,$$

and $x_i = \deg(v_i) - 1$ for $i = 1, 2, \ldots, k$.

This is, in fact, a direct consequence of Proposition 2.2.1, as $W(T)$ can be represented as

$$(n - k)(n - 1) + \sum_{i=1}^{k-1} \left(\left(\sum_{j=1}^{i}(\deg(v_j) - 1) + 1 \right) \cdot \left(\sum_{j=i+1}^{k} (\deg(v_j) - 1) + 1 \right) \right)$$

where the first term corresponds to the contribution from all pendant edges and the second term corresponds to the contribution from all internal edges. The formula (2.4) follows from simple algebra.

Such a formulation turns our problem into a quadratic assignment problem that seeks to maximize the function $q(\mathbf{x})$ where $\mathbf{x}^t$ is a permutation of **b** for a given degree sequence. Although such problems are NP-hard in general, a polynomial time algorithm was found in [14]. The main tool used in the construction of that algorithm is, in addition to formula (2.4), the following so called ∨-*property*.

Theorem 2.4.2 *Following the above notations, among all trees with a given degree sequence, the Wiener index is maximized by a caterpillar satisfying*

$$x_1 \geq x_2 \geq \ldots \geq x_i \leq x_{i+1} \leq \ldots \leq x_k$$

for some $1 \leq i \leq k$.

Proof:

Suppose, for contradiction, that the ∨-property does not hold in a caterpillar with given degree sequence that maximizes the Wiener index. Then, we have

$$x_{i_0-1} < x_{i_0} > x_{i_0+1}$$

for some i_0.

Assume, without loss of generality, that

$$n_{v_{i_0} v_{i_0-1}}(v_{i_0-1}) \geq n_{v_{i_0} v_{i_0+1}}(v_{i_0+1}). \tag{2.5}$$

We now consider the tree T' obtained from T by interchanging the degrees (and hence all corresponding pendant edges) of v_{i_0} and v_{i_0+1} (Figure 2.10).

Note that from T to T' we are simply moving $y = x_{i_0} - x_{i_0+1}$ pendant edges from v_{i_0} to v_{i_0+1}. In this process, only the distances between the corresponding y leaves and the rest of the vertices changed. These include:

- at least $n_{v_{i_0} v_{i_0-1}}(v_{i_0-1}) + 1$ vertices whose distances from these leaves increased by 1;

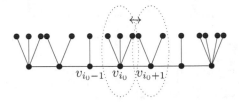

FIGURE 2.10
A caterpillar T with $x_{i_0-1} < x_{i_0} < x_{i_0+1}$.

- exactly $n_{v_{i_0}v_{i_0+1}}(v_{i_0+1})$ vertices whose distances from these leaves decreased by 1.

  Thus, by (2.5) we have

$$W(T') - W(T) \geq y \cdot \left(\left(n_{v_{i_0}v_{i_0-1}}(v_{i_0-1}) + 1 \right) - \left(n_{v_{i_0}v_{i_0+1}}(v_{i_0+1}) \right) \right) > 0,$$

a contradiction. $\square$

  Assuming

$$b_1 > b_2 > b_3 > \cdots > b_k$$

in our decremented degree sequence, full characterization of the extremal caterpillars that maximize the Wiener index can be achieved for small values of k. Here, we simply use a permutation of the decremented degree sequence to denote the corresponding extremal caterpillar:

- for $k = 4$, (b_1, b_4, b_3, b_2) maximizes the Wiener index;
- for $k = 5$, there are three cases:
    - $(b_1, b_5, b_4, b_3, b_2)$ maximizes if $b_1 - b_2 - b_3 > 0$,
    - $(b_1, b_4, b_5, b_3, b_2)$ maximizes if $b_1 - b_2 - b_3 < 0$, and
    - both $(b_1, b_5, b_4, b_3, b_2)$ and $(b_1, b_4, b_5, b_3, b_2)$ maximize if $b_1 - b_2 - b_3 = 0$.
- for $k = 6$ there are 12 cases.

  See [97] for details. In addition to the partial characterizations of the extremal caterpillars with a given degree sequence that maximize the Wiener index, it is of practical interest to consider the same question restricted to the "chemical trees" whose vertices are of degree ≤ 4, in which case $1 \leq b_i \leq 3$.
  More generally, if

$$\{b_1, b_2, \ldots, b_k\} = \{\underbrace{a_s, \ldots, a_s}_{m_s}, \underbrace{a_{s-1}, \ldots, a_{s-1}}_{m_{s-1}}, \ldots, \underbrace{a_1, \ldots, a_1}_{m_1}\}$$

with $a_s > a_{s-1} > \ldots > a_1$, then in the extremal tree we must have

$$\mathbf{x} = \{\underbrace{a_s, \ldots, a_s}_{l_s}, \underbrace{a_{s-1}, \ldots, a_{s-1}}_{l_{s-1}}, \ldots, \underbrace{a_1, \ldots, a_1}_{m_1}, \ldots, \underbrace{a_{s-1}, \ldots, a_{s-1}}_{r_{s-1}}, \underbrace{a_s, \ldots, a_s}_{r_s}\}$$

from the $\vee$-property, where $l_i + r_i = m_i$ for $i = 2, 3, \ldots, s$. Consequently, we have

$$q(\mathbf{x}) = \sum X_{\alpha,\beta} \qquad (2.6)$$

where α or β (might be the same) corresponds to one of the sequences of equal entries such as $\underbrace{\{a_s, \ldots, a_s\}}_{l_s}$ or $\underbrace{\{a_1, \ldots, a_1\}}_{m_1}$.

For instance, if $\alpha = \underbrace{\{a_s, \ldots, a_s\}}_{l_s}$ and $\beta = \underbrace{\{a_{s-1}, \ldots, a_{s-1}\}}_{r_{s-1}}$,

$$X_{\alpha,\beta} = a_s a_{s-1} \sum_{i=0}^{l_s-1} \sum_{j=x+1+i}^{x+r_{s-1}+i} j = \frac{a_s a_{s-1}}{2} l_s r_{s-1}(2x + l_s + r_{s-1})$$

where x is the *distance* between α and β, $l_{s-1} + \ldots + l_2 + m_1 + r_2 + \ldots + r_{s-2}$. In the case $\alpha = \beta = \underbrace{\{a_s, \ldots, a_s\}}_{l_s}$,

$$X_{\alpha,\alpha} = \frac{a_s^2}{6} l_s (l_s - 1)(l_s + 1).$$

Applying (2.6) to $\mathbf{x} = \{\underbrace{3, \ldots, 3}_{l_3}, \underbrace{2, \ldots, 2}_{l_2}, \underbrace{1, \ldots, 1}_{m_1}, \underbrace{2, \ldots, 2}_{r_2}, \underbrace{3, \ldots, 3}_{r_3}\}$ (i.e., the chemical trees), it follows from simple algebra that the Wiener index is maximized when the 3s and 2s are evenly distributed on both ends. In fact, replacing $\{3, 2, 1\}$ with $\{a, b, c\}$ for any $a > b > c$ yields the same conclusion.

2.5 The Wiener index of trees with a given segment sequence

As mentioned earlier, fixing the segment sequence introduces a unique and interesting restriction on trees. In this section, we consider the trees with a given segment sequence that maximize or minimize the Wiener index.

2.5.1 The minimum Wiener index in trees with a given segment sequence

For a given segment sequence $(l_1, l_2, \ldots, l_m)$, recall that the *starlike tree* $S(l_1, l_2, \ldots, l_m)$ is the tree with exactly one vertex of degree ≥ 3 formed by identifying one end of each of the m segments. It was first shown in [70] that $S(l_1, l_2, \ldots, l_m)$ minimizes the Wiener index among all trees with segment sequence $(l_1, l_2, \ldots, l_m)$. To illustrate different approaches in chemical graph theory we provide two proofs for this simple result; one has a "computational flavor", while the other focuses more on the structural properties.

Theorem 2.5.1 *Among trees with a given segment sequence*

$$(l_1, l_2, \ldots, l_m),$$

the starlike tree $S(l_1, l_2, \ldots, l_m)$ *minimizes the Wiener index.*

Proof:

- The first proof is rather similar to that for the extremality of the starlike tree with respect to the terminal Wiener index, based on the expression

$$W(T) = \sum_{uv \in E(T)} n_{uv}(v) \cdot n_{uv}(u)$$

from Proposition 2.2.1.

Note that for a segment of length ℓ, at most one end of it is a leaf, and hence the contribution $n_{uv}(v) \cdot n_{uv}(u)$ from each edge to the above formula is at least

$$1 \cdot (n-1) + 2 \cdot (n-2) + \ldots + \ell \cdot (n-\ell)$$

where n is the number of vertices in T. This "least" contribution can be achieved if and only if every segment has one end being a leaf, which can only happen when the tree has exactly one branching vertex.

- As a second approach, we consider (for contradiction) a tree T that minimizes the Wiener index with at least two branching vertices.

Let u and w be two of the branching vertices such that the path connecting them is a segment (i.e., there are no branching vertices "between" them). Now we define S to be the component containing u (with root u) in $T - P(u, w)$, and R to be $T - (S - u)$.

Following the setup in Lemma 2.2.1, it is easy to verify

$$d_R(w) < d_R(u).$$

We leave this as an exercise to interested readers. Consequently, Lemma 2.2.1 implies that the tree T' obtained by identifying the root of S with w of R has smaller Wiener index than T (considered as identifying the root of S with u of R), a contradiction.

$\square$

2.5.2 The maximum Wiener index in trees with a given segment sequence

Recall that a *quasi-caterpillar* is a tree all of whose branching vertices lie on a path (Figure 1.8). We first show that quasi-caterpillars maximize the Wiener index. This was conjectured in [70] and proved in [5], whose proof we provide here.

Theorem 2.5.2 *If a tree T maximizes the Wiener index among all trees with the same segment sequence, then it must be a quasi-caterpillar.*

Proof:

Let T be an extremal tree with the given segment sequence that maximizes the Wiener index, and let $P = P(v_0, v_k)$ be a path that contains the greatest number of segments. Then both v_0 and v_k must be leaves. We now label the branching vertices on P by $v_1, v_2, \ldots, v_{k-1}$ in the order of their distances from v_0.

For each $1 \leq i \leq k - 1$, since v_i is a branching vertex, it has neighbors $v_{i1}, \ldots, v_{il_i}$ not on P. Further let T_{ij} $(1 \leq j \leq l_i)$ denote the component containing v_{ij} in $T - v_i v_{ij}$.

In every T_{ij}, denote by u_{ij} the branching vertex (or leaf if there is no branching vertex) closest to v_i and by S_{ij} the component containing u_{ij} in $T - E(P(v_i, u_{ij}))$. Figure 2.11 provides a detailed illustration for these notations.

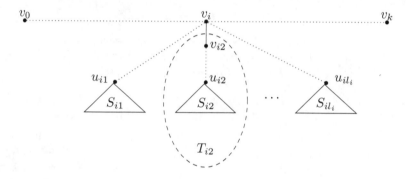

FIGURE 2.11
The labeling of T.

Supposing for contradiction that T is not a quasi-caterpillar, we must have some S_{ij} with more than one vertex. Let $S = S_{i_0 j_0}$ be the one with the greatest number of vertices among all S_{ij} $(1 \leq i \leq k, 1 \leq j \leq l_i)$, let $T_{\leq i_0}$ denote the component containing v_{i_0} in $T - E(P(v_{i_0}, v_{i_0+1}))$ and $T_{>i_0}$ the component containing v_{i_0+1} in $T - E(P(v_{i_0}, v_{i_0+1}))$. Similarly for $T_{<i_0}$ and $T_{\geq i_0}$.

Assume now, without loss of generality, that

$$|T_{<i_0}| \geq |T_{>i_0}|. \tag{2.7}$$

We may also assume

$$|S| > |S_{ij}|$$

for all $i > i_0$ and all j. This is because we can always let $|S_{ij}| = |S|$ for some maximal i, while still maintaining (2.7).

Now the path $P = P(v_0, v_k)$, according to the above discussion, is one containing the greatest number of segments with $i_0 \neq k-1$ (i.e., v_{i_0} cannot be the last branching vertex, as then there would be a path through $u_{i_0 j_0}$ rather than v_k that contains more segments). Therefore v_{i_0+1} is still a branching vertex.

We will now study the subtree $T_{i_0+1,1}$ obtained from adding to the subtree $S' = S_{i_0+1,1}$ the path from v_{i_0+1} to $u_{i_0+1,1}$ (Figure 2.12).

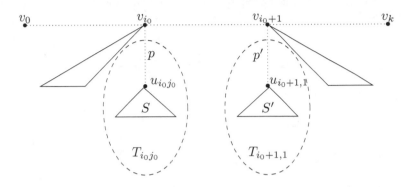

FIGURE 2.12
The branches that are switched.

Let p and p' be the lengths of the paths $P(v_{i_0}, u_{i_0 j_0})$ and $P(v_{i_0+1}, u_{i_0+1,1})$, respectively. We have the following cases:

1. If $p \geq p'$, let T' be obtained from T by "switching" the subtrees $T_{i_0 j_0}$ and $T_{(i_0+1),1}$. Or equivalently,

$$T' = T \quad v_{i_0} v_{i_0 j_0} \quad v_{i_0+1} v_{(i_0+1),j} + v_{i_0} v_{(i_0+1),j_0} + v_{i_0+1} v_{i_0};$$

2. If $p < p'$, let T' be obtained from T by "switching" the subtrees S and S'. That is, to replace $T_{i_0 j_0}$ with (p, S') and $T_{(i_0+1),1}$ with $(p', S_{i_0 j_0})$.

In both cases, it is not hard to check that the segment sequence stays the same from T to T'. We will now examine, in detail, how the distances between vertices change from T to T'.

- In the first case, the distances change as follows:
    - the distance between any vertex in $T_{i_0 j_0}$ and any vertex in $T_{\leq i_0} - T_{i_0 j_0}$ increases by $d(v_{i_0}, v_{i_0+1})$;
    - the distance between any vertex in $T_{i_0 j_0}$ and any vertex in $T_{>i_0} - T_{(i_0+1),1}$ decreases by $d(v_{i_0}, v_{i_0+1})$;
    - the distance between any vertex in $T_{(i_0+1),1}$ and any vertex in $T_{\leq i_0} - T_{i_0 j_0}$ decreases by $d(v_{i_0}, v_{i_0+1})$;
    - the distance between any vertex in $T_{(i_0+1),1}$ and any vertex in $T_{>i_0} - T_{(i_0+1),1}$ increases by $d(v_{i_0}, v_{i_0+1})$;

- the distances between vertices of $T_{i_0 j_0}$ and the vertices on the segment between v_{i_0} and v_{i_0+1} change, but the total contribution to the Wiener index remains the same; the same is true for $T_{i_0+1,1}$.

- all distances between other pairs of vertices stay the same.

Consequently, the total change (of the value of the Wiener index) in this case is

$$W(T') - W(T)$$
$$= d(v_{i_0}, v_{i_0+1})\big(|T_{i_0 j_0}| - |T_{(i_0+1),1}|\big) \cdot \big(|T_{\leq i_0} - T_{i_0 j_0}| - |T_{>i_0} - T_{(i_0+1),1}|\big).$$

Note that

$$|T_{\leq i_0} - T_{i_0 j_0}| > |T_{<i_0}| \geq |T_{>i_0}| > |T_{>i_0} - T_{i_0+1,1}|,$$

and by the fact $|S| > |S'|$ from the assumption $S = S_{i_0 j_0}$, we have

$$|T_{i_0 j_0}| > |T_{i_0+1,1}|.$$

Thus $W(T') > W(T)$, a contradiction.

- In the second case, we only need to consider the distances between vertices in S and S' and the rest of the tree. Through similar reasoning to the first case, we have

$$W(T') - W(T) = \Big(d(v_{i_0}, v_{i_0+1})\big(|T_{\leq i_0} - T_{i_0 j_0}| - |T_{>i_0} - T_{i_0+1,1}|\big)$$
$$+ (p' - p)\big(|T_{\leq i_0} - T_{i_0 j_0}| + |T_{>i_0} - T_{i_0+1,1}|\big)\Big)\big(|S| - |S'|\big),$$

yielding another contradiction.

In both cases, we see that T cannot be extremal, hence completing the proof by contradiction. $\qquad\square$

2.5.3   Further characterization of extremal quasi-caterpillars

For trees with a given degree sequence, it is relatively easy to show that the maximum Wiener index is obtained by a caterpillar (see the proof of Proposition 2.6.1), but specific characterizations of the extremal caterpillar turned out to be a difficult problem. In the case of maximizing the Wiener index among trees with a given segment sequence, it is also an interesting and challenging problem to identify further characteristics of the extremal quasi-caterpillar that maximizes the Wiener index.

Similar to the backbone of a caterpillar, we will let the longest path of a quasi-caterpillar containing all the branching vertices be called the *backbone*. It is easy to see that each of the other segments that are not on the backbone connects a leaf with a branching vertex. We will call them *pendant segments*.

Given a segment sequence $(l_1, l_2, \ldots, l_m)$, Theorem 2.5.2 states that the maximum Wiener index can only be attained by a quasi-caterpillar. In what follows we present some further characteristics of extremal quasi-caterpillars, also established in [5]. The technical details are somewhat similar to the proof of Theorem 2.5.2, some of which we leave to interested readers as exercises.

Theorem 2.5.3 *In a quasi-caterpillar that maximizes the Wiener index among trees with segment sequence $(l_1, l_2, \ldots, l_m)$ we must have the following:*

- *If the number of segments is odd, all branching vertices have degree exactly 3;*

- *If the number of segments is even, all but one branching vertices have degree 3. The only exception must be a last branching vertex of degree 4, which must be the first (or last) branching vertex on the backbone.*

This also means that the number of segments on the backbone is $k = \lfloor (m + 1)/2 \rfloor$ and the number of pendant segments is $k' = \lceil (m-1)/2 \rceil$.

Proof:

Similar to before, we start with labeling the backbone as a path $P(v_0, v_k)$ between leaves v_0 and v_k with branching vertices $v_1, v_2, \ldots, v_{k-1}$ (in the order of their distances from v_0).

First we show that in the extremal quasi-caterpillar, no branching vertex is of degree greater than 4.

Supposing (for contradiction) otherwise, let v_i be of degree at least 5 with neighbors $v_{i1}, v_{i2}, v_{i3}, \ldots$ not on the path $P(v_0, v_k)$. Again we let $T_{<i}$ ($T_{>i}$) denote the component containing v_{i-1} (v_{i+1}) in $T - E(P(v_{i-1}, v_{i+1}))$ as before, and let $T_{\leq i} = T_{<i+1}$ and $T_{\geq i} = T_{>i-1}$ also be defined as in the proof of Theorem 2.5.2. Lastly, let T_{i1}, T_{i2}, T_{i3} be the pendant segments at v_i containing v_{i1}, v_{i2}, v_{i3}, respectively.

We may assume, without loss of generality, that

$$|T_{<i}| \geq |T_{>i}|,$$

and hence

$$|T_{\leq i} - T_{i1} - T_{i2}| > |T_{>i}|.$$

Consider now T', the tree obtained from T by detaching T_{i1} and T_{i2} from v_i and reattaching them to v_{i+1}. It is easy to check that T' has the same segment sequence as T, even in the special case $i = k - 1$. Similar discussions to those in the proof of Theorem 2.5.2 show that

$$W(T') - W(T) = d(v_i, v_{i+1}) \left(|T_{i1}| + |T_{i2}| \right) \cdot \left(|T_{\leq i} - T_{i1} - T_{i2}| - |T_{>i_0}| \right) > 0,$$

a contradiction. Thus all branching vertices are of degree 3 or 4.

Now, for a vertex v_i of degree 4, we may repeat the above arguments with moving only one segment instead of two to obtain the tree T'. A contradiction follows except for the cases when $v_i = v_1$ or $v_i = v_{k-1}$; in those cases where $v_i = v_1$ or $v_i = v_{k-1}$, the described operation would move a single segment to the end of the backbone and change the segment sequence. Thus, the only branching vertices that could possibly have degree 4 are v_1 and v_{k-1}.

Lastly, we consider the scenario when both v_1 and v_{k-1} are of degree 4.

Let $S = T_{11}$ and $S' = T_{k-1,1}$ be two segments attached to v_1 and v_{k-1}, respectively, and let R be obtained from T by removing these two segments (Figure 2.13). We may assume, without loss of generality, that

$$\sum_{v \in V(R)} d(v_k, v) \leq \sum_{v \in V(R)} d(v_0, v).$$

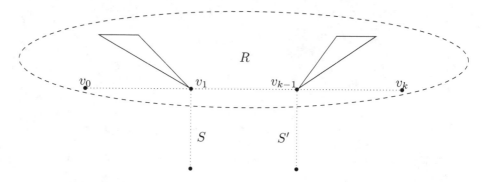

FIGURE 2.13
The segments S and S' and the rest of the tree (denoted R).

Construct T' from T by removing both S and S' and reattaching them to v_0. Again, it is easy to check that the segment sequence stays the same from T to T'. The change in the value of the Wiener index from T to T' is discussed through the distances between various pairs of vertices as before:

- the total distance between any two vertices in S, any two vertices in S', or any two vertices in R does not change;

- the total distance between vertices in S and vertices in S' decreases by $d(v_1, v_{k-1})|S||S'|$;

- the total distance between vertices in S and $P(v_0, v_1)$ does not change, while the total distance between vertices in S and the rest of R increases by $d(v_0, v_1)(|R| - d(v_0, v_1) - 1)|S|$;

- if S' is moved to v_k, the total distance between vertices in S' and R increases by $d(v_{k-1}, v_k)(|R| - d(v_{k-1}, v_k) - 1)|S'|$ as before;

- moving S' further to v_0 changes the total distance further by

$$|S'|\Big(\sum_{v \in V(R)} d(v_0, v) - \sum_{v \in V(R)} d(v_k, v) \Big).$$

Recall that the backbone is the longest path that contains all the branching vertices, which implies $|S| \leq d(v_0, v_1)$ and $|S'| \leq d(v_{k-1}, v_k)$.

It is also easy to see that

$$|R| > d(v_0, v_k) + 1 = d(v_0, v_1) + d(v_1, v_{k-1}) + d(v_{k-1}, v_k) + 1.$$

Consequently, through direct computation, we have that

$$
\begin{aligned}
W(T') - W(T) = {}& |S'| \Big(\sum_{v \in V(R)} d(v_0, v) - \sum_{v \in V(R)} d(v_{k+1}, v) \Big) \\
& + d(v_{k-1}, v_k)\big(|R| - d(v_{k-1}, v_k) - 1\big)|S'| \\
& + d(v_0, v_1)\big(|R| - d(v_0, v_1) - 1\big)|S| - d(v_1, v_{k-1})|S||S'| \\
> {}& d(v_{k-1}, v_k)\big(d(v_0, v_1) + d(v_1, v_{k-1})\big)|S'| \\
& + d(v_0, v_1)\big(d(v_1, v_{k-1}) + d(v_{k-1}, v_k)\big)|S| - d(v_1, v_{k-1})|S||S'| \\
> {}& d(v_1, v_{k-1})\big(|S'|^2 + |S|^2 - |S||S'|\big) \\
> {}& 0.
\end{aligned}
$$

This is a contradiction. Hence, there can be at most one vertex of degree 4. Should such a vertex exist, the number of segments must be even and this vertex has to be either v_1 or v_{k-1}. □

Now that we have established that all branching vertices are of small (as small as possible, in fact) degrees, we go on to partially characterize the arrangement of segments along the backbone in an extremal quasi-caterpillar. The following is, in some sense, similar to the ∨-property that we established for the extremal caterpillars with a given degree sequence.

Theorem 2.5.4 *Let T be a quasi-caterpillar that maximizes the Wiener index among trees with segment sequence $(l_1, l_2, \ldots, l_m)$. In T, the lengths of the segments on the backbone, listed from one end to the other, form a unimodal sequence $r_1, r_2, \ldots, r_k$, i.e.,*

$$r_1 \leq r_2 \leq \cdots \leq r_j \geq \cdots \geq r_k$$

for some $j \in \{1, 2, \ldots, k\}$.

Proof:

Following the same notations, consider now the segments

$$P(v_0, v_1), P(v_1, v_2), \ldots, P(v_{k-1}, v_k)$$

on the backbone of T. For convenience let $r_1, r_2, \ldots, r_k$ be the lengths of these segments, and let M be the maximum length of a backbone segment. If all r_i's are equal, there is nothing to prove. Assume that is not the case and further let j be the smallest index such that

$$r_j = d(v_{j-1}, v_j) = M > r_{j+1} = d(v_j, v_{j+1}).$$

Note that such an index always exists (if necessary, after reversing the backbone and all labels).

Using similar notations as before, we let $T_{\leq j-1}$, T_j and $T_{\geq j+1}$ denote the components containing v_{j-1}, v_j and v_{j+1}, respectively, in $T - E(P(v_{j-1}, v_{j+1}))$ (Figure 2.14).

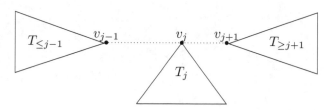

FIGURE 2.14
The subtrees $T_{\leq j-1}$, T_j and $T_{\geq j+1}$.

If $|T_{\leq j-1}| < |T_{\geq j+1}|$, let T' be obtained from T by "interchanging" $T_{\leq j-1}$ and $T_{\geq j+1}$. Then,

$$W(T') - W(T) = (|T_j| - 1)(r_j - r_{j+1})\big(|T_{\geq j+1}| - |T_{\leq j-1}|\big) > 0,$$

a contradiction.

Thus, we must have

$$|T_{\leq j-1}| \geq |T_{\geq j+1}|$$

and consequently,

$$|T_{\leq i-1}| > |T_{\leq j-1}| \geq |T_{\geq j+1}| > |T_{\geq i+1}|$$

for all $i > j$. This, in turn, implies that $r_i \geq r_{i+1}$ through the same argument. Thus, we have

$$r_j \geq r_{j+1} \geq \cdots .$$

The proof for $r_1 \leq \cdots \leq r_j$ is similar.   □

We now show one more property of the extremal quasi-caterpillars with a given segment sequence, this time on the pendant segments, similar to the $\vee$-property of extremal caterpillars with a given degree sequence.

Theorem 2.5.5 *Let T be a quasi-caterpillar that maximizes the Wiener index among trees with segment sequence $(l_1, l_2, \ldots, l_m)$. In T, the lengths of the* *pendant segments, starting from one end of the backbone towards the other, form a sequence of values* $s_1, s_2, \ldots, s_{k'}$ *such that*

$$s_1 \geq s_2 \geq \cdots \geq s_{j'} \leq \cdots \leq s_{k'}$$

for some $j' \in \{1, 2, \ldots, k'\}$.

Proof:

We will assume, without loss of generality, that the number of segments in T is odd. The even case can be argued in exactly the same way.

Then, by Theorem 2.5.3, all branching vertices are of degree 3. Let S_i denote the pendant segment at v_i for all $1 \leq i \leq k' = k - 1$ and let s_i denote its length. Further, let μ be the minimum length of all pendant segments.

Similar to the proof of Theorem 2.5.4, we will assume that not all of these segments are of the same length. Then there exists (possibly after reversing the order) a smallest j' such that

$$s_{j'} = \mu < s_{j'+1}.$$

Also let $T_{\leq j'}$ and $T_{\geq j'+1}$ denote the components containing $v_{j'}$ and $v_{j'+1}$, respectively, in $T - E(P(v_{j'}, v_{j'} + 1))$.

Now "interchanging" $S_{j'}$ and $S_{j'+1}$ in T results in a tree T' where

$$W(T') - W(T) = d(v_{j'}, v_{j'+1})(s_{j'+1} - s_{j'})\big(|T_{\geq j'+1} - S_{j'+1}| - |T_{\leq j'} - S_{j'}|\big),$$

implying that we must have

$$|T_{\leq j'} - S_{j'}| \geq |T_{\geq j'+1} - S_{j'+1}|.$$

Thus,

$$|T_{\leq i} - S_i| \geq |T_{\leq j'}| > |T_{\leq j'} - S_{j'}| \geq |T_{\geq j'+1} - S_{j'+1}| \geq |T_{\geq i+1}| > |T_{\geq i+1} - S_{i+1}|$$

for any $i > j'$, which implies that $s_{i+1} \geq s_i$ by the same argument. It follows that $s_{j'} \leq s_{j'+1} \leq \cdots$.

Similarly, we also have $s_1 \geq \cdots \geq s_{j'}$. □

2.5.4 Trees with a given number of segments

Also in [70], the tree with minimal Wiener index among all trees with a given number of segments is characterized. For trees with n vertices and m segments, the extremal tree that minimizes the Wiener index must be a starlike tree (as established earlier in this section) and consequently must have exactly m leaves. The question is then turned into minimizing the Wiener index with given numbers of vertices and leaves, resulting in a starlike tree on m segments such that the difference between the lengths of any pair of segments is at most

1 (i.e., evenly distributed). Such extremal structures are special cases of more general results that we will discuss later.

In the current section, once again we are interested in the analogous question for the maximum Wiener index (among trees with a given number of segments), conjectured in [70] and shown in [5].

First, we need to introduce the corresponding extremal structures. For given n and m, we define trees $O(n,m)$ (for odd m) and $E(n,m)$ (for even m), respectively.

- The graph $O(n,m)$ is obtained from a path $v_0, v_1 \ldots, v_\ell$ of length $\ell = n - \frac{m+1}{2}$ by attaching a total of $\frac{m-1}{2}$ leaves to vertices $v_1, v_2, \ldots, v_{\lfloor (m-1)/4 \rfloor}$ and $v_{\ell-1}, v_{\ell-2}, \ldots, v_{\ell-\lceil (m-1)/4 \rceil}$, see Figure 2.15 (left) for the case $n = 11$, $m = 7$. Note that $O(n,m)$ has exactly m segments.

- Likewise, $E(n,m)$ is a tree with n vertices and m segments obtained from a path $v_0, v_1 \ldots, v_\ell$ of length $\ell = n - \frac{m}{2} - 1$ by attaching a total of $\frac{m}{2}$ leaves to vertices $v_1, v_2, \ldots, v_{\lfloor (m-2)/4 \rfloor}$ and $v_{\ell-1}, v_{\ell-2}, \ldots, v_{\ell-\lceil (m-2)/4 \rceil}$, where two leaves are attached to vertex v_1 (so that it becomes the only vertex of degree 4), see Figure 2.15 (right) for the case $n = 11$, $m = 8$.

FIGURE 2.15
The trees $O(11, 7)$ and $E(11, 8)$.

With Theorems 2.5.3, 2.5.4, and 2.5.5 established in the previous section, it is now easy to show that $O(n,m)$ (if m is odd) and $E(n,m)$ (if m is even) always maximize the Wiener index among all trees of order n with m segments. The following proof was given in [5].

Theorem 2.5.6 *Among all trees of order n with m segments, $O(n,m)$ $(E(n,m))$ maximizes the Wiener index if m is odd (even).*

Proof:

We will only consider the case of odd m and leave the other case as an exercise.

Let T be the extremal tree, with the given numbers of vertices and segments, that maximizes the Wiener index. Then, by Theorems 2.5.2 and 2.5.3, T has to be a quasi-caterpillar, and every branching vertex has degree 3.

As before, let the path $P(v_0, v_k)$ be the backbone of the quasi-caterpillar with branching vertices $v_1, v_2, \ldots, v_{k-1}$. Note that the total number of segments is $m = 2k - 1$.

Moreover, let a and b be the lengths of $P(v_0, v_1)$ and the other pendant segment ending at v_1. We let T' be the tree obtained from removing those two segments (including v_1) from T.

Suppose that $\min\{a, b\} > 1$. It is an easy exercise to show that replacing the two segments by segments of lengths 1 and $a + b - 1$ will increase the Wiener index by $(a - 1)(b - 1)|T'|$, and hence a contradiction to the choice of T.

Thus, the pendant segment at v_1 has to have length 1. Similarly, the pendant segment at v_{k-1} also has to be of length 1. By Theorem 2.5.5, all pendant segments must have length 1. This implies that T is a caterpillar.

From the $\vee$-property of the extremal caterpillar that we established in the previous section, we can conclude that the degrees of the internal vertices along the backbone have to be decreasing at first, then increasing, i.e., the sequence of degrees has to be of the form

$$3, 3, \ldots, 3, 2, 2, \ldots, 2, 2, 3, 3, \ldots, 3.$$

To finish our proof, it only remains to show that the number of vertices of degree 3 on the two sides differ by at most 1. For this purpose, we will relabel the vertices on the backbone as

$$u_0 = v_0, u_1, u_2, \ldots, u_{n-k} = v_k.$$

Note that this includes all vertices instead of just the branching vertices. Assume now that there is a leaf attached to $u_1, u_2, \ldots, u_x$ and $u_{n-k-1}, u_{n-k-2}, \ldots, u_{n-k-y}$, where $x + y = k - 1$.

If $k - 1 = n - k - 1$ (equivalently, $n = 2k = m + 1$), then all vertices on the backbone have to have degree 3 and we are done.

Otherwise, assume that $|x - y| > 1$ and, without loss of generality, $x > y + 1$. Considering the tree T' obtained from moving one leaf from u_x to $u_{n-k-y-1}$, it is again an easy exercise to show that

$$W(T') - W(T) = 2(x - y - 1)(n - 2k) > 0,$$

and we reach yet another contradiction. Thus, $|x - y| \leq 1$, implying that T is isomorphic to $O(n, m)$. □

2.6   General approaches

So far we have repeatedly mentioned more general extremal results regarding distance-based indices. This section is devoted to such discussions. We start with defining a topological index of the form

$$W_f(T) = \sum_{\{u,v\} \subseteq V(T)} f(d(u, v))$$

for some function f. With different choices of f, $W_f(\cdot)$ defines various distance-based indices such as the Wiener index and its generalization

$$W_\alpha(G) = \sum_{\{v,w\} \subseteq V(G)} d(v,w)^\alpha,$$

the *hyper-Wiener index* [64]

$$WW(G) = \sum_{\{v,w\} \subseteq V(G)} \binom{d(v,w)+1}{2},$$

the *Harary index* [88]

$$H(T) = \sum_{\{u,v\} \subseteq V(T)} \frac{1}{d(u,v)},$$

and many more.

We will first establish the extremality of the greedy trees and caterpillars among trees with a given degree sequence, with respect to $W_f(\cdot)$ for functions f under specific restrictions. Then, we will compare extremal trees of different degree sequences. Such comparisons lead to immediate consequences in which extremal structures with respect to various distance-based indices are identified in different classes of trees.

2.6.1   Caterpillars

We start with the problem of maximizing the Wiener index and its generalizations among trees of a given degree sequence. Similar to Section 2.5.2, although it is not possible to fully characterize the solution, the problem can be reduced to the study of caterpillars. Here, as well as in the rest of this section, we follow the proofs introduced in [94].

Proposition 2.6.1 *Let $f(x)$ be a strictly increasing and convex function (i.e., the increments $f(x+1) - f(x)$ are non-decreasing). If T is a tree that maximizes*

$$W_f(T) = \sum_{\{u,v\} \subseteq V(T)} f(d(u,v))$$

among all trees with degree sequence $(d_1, \ldots, d_n)$, then T must be a caterpillar.

Proof:

Let P be a longest path of this extremal tree T, let the vertices on P be denoted $x_1, x_2, \ldots, x_\ell$, and suppose (for contradiction) that T is not a caterpillar. Then, we must have $\ell \geq 4$ and there exists an x_k, $2 < k < \ell - 1$, such that x_k has a non-leaf neighbor y that is not on P. Let

$$N(y) = \{x_k, z_1, \ldots, z_s\},$$

for some $s \geq 1$, be the neighbors of y. We also use T_1, T_2 and T_3 to denote the components containing x_k, x_{k+1} and y, respectively, in $T - x_k x_{k+1} - x_k y$ (Figure 2.16). Without loss of generality, we can further assume that $|V(T_1)| \geq |V(T_2)| \geq 2$.

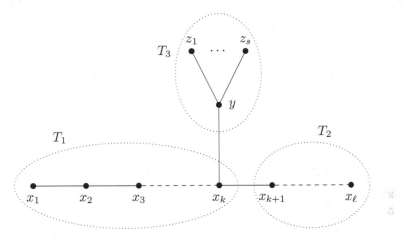

FIGURE 2.16
T_1, T_2 and T_3 in an extremal tree T that is not a caterpillar.

Let T' be obtained from T by replacing each edge yz_i of T by the new edge $x_\ell z_i$, for $i = 1, \ldots, s$. It is easy to see that T' and T have the same degree sequence.

Consider the distance between two vertices u and v in T and T'. Note that $d_{T'}(u, v) \neq d_T(u, v)$ only when $u \in V(T_3) \setminus \{y\}$ and $v \in V(T_1) \cup V(T_2) \cup \{y\}$ (or vice versa). Since the contributions of y and x_ℓ to $W_f(\cdot)$ cancel, it suffices to consider $v \in V(T_1)$ and $v \in V(T_2) \setminus \{x_\ell\}$.

Consequently we have

$W_f(T') - W_f(T)$

$$
= \sum_{u \in V(T_3) \setminus \{y\}} \left[\sum_{v \in V(T_1)} \left(f(d_{T'}(u, v)) - f(d_T(u, v)) \right) \right.
$$

$$
+ \sum_{v \in V(T_2) \setminus \{x_\ell\}} \left. \left(f(d_{T'}(u, v)) - f(d_T(u, v)) \right) \right]
$$

$$
= \sum_{u \in V(T_3) \setminus \{y\}} \left[\sum_{v \in V(T_1)} \left(f(d_T(u, y) + d_T(x_k, x_\ell) + d_T(x_k, v)) \right. \right.
$$

$$
- f(d_T(u, y) + d_T(x_k, v) + 1)) + \sum_{v \in V(T_2) \setminus \{x_\ell\}} \left(f(d_T(u, y) + d_T(x_\ell, v)) \right.
$$

$$
\left. \left. - f(d_T(u, y) + d_T(x_k, v) + 1)) \right].
$$

Since P is a longest path in T, we have

$$d_T(x_k, v) \leq d_T(x_k, x_\ell)$$

for all $v \in V(T_2) \setminus \{x_\ell\}$ and $d_T(x_k, x_\ell) \geq 2$. We also point out the simple facts that

$$d_T(x_\ell, x_{\ell-1}) = 1 \text{ and } d_T(x_\ell, v) \geq 2$$

for all $v \in V(T_2) \setminus \{x_\ell, x_{\ell-1}\}$.

The contribution from each $u \in V(T_3) \setminus \{y\}$ to $W_f(T') - W_f(T)$ is

$$\sum_{v \in V(T_1)} \left[f(d_T(u, y) + d_T(x_k, x_\ell) + d_T(x_k, v)) \right.$$

$$- f(d_T(u, y) + d_T(x_k, v) + 1) \Big]$$

$$+ \sum_{v \in V(T_2) \setminus \{x_{\ell-1}, x_\ell\}} \left[f(d_T(u, y) + d(x_\ell, v)) - f(d_T(u, y) + d_T(x_k, v) + 1) \right]$$

$$+ f(d_T(u, y) + 1) - f(d_T(u, y) + d_T(x_k, x_\ell)).$$

In view of the convexity of f and the aforementioned inequalities, this is at least

$$\sum_{v \in V(T_1) \setminus \{x_k\}} \left[f(d_T(u, y) + d_T(x_k, x_\ell) + 1) - f(d_T(u, y) + 2) \right]$$

$$+ f(d_T(u, y) + d_T(x_k, x_\ell)) - f(d_T(u, y) + 1)$$

$$+ \sum_{v \in V(T_2) \setminus \{x_{\ell-1}, x_\ell\}} \left[f(d_T(u, y) + 2) - f(d_T(u, y) + d_T(x_k, x_\ell) + 1) \right]$$

$$+ f(d_T(u, y) + 1) - f(d_T(u, y) + d_T(x_k, x_\ell))$$

$$= \left[f(d_T(u, y) + d_T(x_k, x_\ell) + 1) - f(d_T(u, y) + 2) \right]$$

$$\cdot \left(|V(T_1)| - 1 - |V(T_2)| + 2 \right)$$

$$\geq f(d_T(u, y) + d_T(x_k, x_\ell) + 1) - f(d_T(u, y) + 2) > 0.$$

Consequently, we have $W_f(T') > W_f(T)$, a contradiction. □

Note that we do need the function f to be strictly increasing to claim that the extremal tree must be a caterpillar. In the case that f is only non-decreasing, we obtain the following weaker result. We leave the proof as an exercise.

Proposition 2.6.2 *Let $f(x)$ be a non-decreasing and convex function (i.e., the increments $f(x + 1) - f(x)$ are non-decreasing). Among all trees with degree sequence $(d_1, \ldots, d_n)$ there always exists a caterpillar that achieves the maximum value of*

$$W_f(T) = \sum_{\{u,v\} \subseteq V(T)} f(d(u, v)).$$

Remark 2.6.1 *Proposition 2.6.1 is wrong if f is not convex. Consider, for instance, $f(x) = \sqrt{x}$ and the degree sequence*

$$(20, 20, 20, 3, 1, 1, \ldots, 1).$$

There are only two non-isomorphic caterpillars in this case, and the value of $W_f(\cdot)$ for these two caterpillars is

$$858 + 573\sqrt{2} + 760\sqrt{3} + 38\sqrt{5} \approx 3069.67$$

and

$$858 + 573\sqrt{2} + 437\sqrt{3} + 361\sqrt{5} \approx 3232.47,$$

respectively. However, it turns out that a tree with a center of degree 3 whose neighbors all have degree 20 (with all the other vertices being leaves) is optimal in this case: for this tree, W_f attains a value of

$$2226 + 573\sqrt{2} + 114\sqrt{3} \approx 3233.8.$$

In the case of the Wiener index ($f(x) = x$), hyper-Wiener index ($f(x) = \binom{x+1}{2}$)), and the generalized Wiener index ($f(x) = x^\alpha$) with $\alpha > 1$, the corresponding functions f are indeed strictly increasing and convex. Thus, Proposition 2.6.1 holds.

Remark 2.6.2 *It is also interesting to point out that, even with the same degree sequence, the extremal tree/caterpillar differs for different functions f. For example, let*

$$(d_1, \ldots, d_7) = (80, 76, 60, 30, 11, 6, 2)$$

be the degree sequence of the internal vertices. Then, the unique extremal caterpillar with respect to the Wiener index is the caterpillar T_1 with backbone vertex degrees in the order

$$(d_{1,1}, \ldots, d_{1,7}) = (d_1, d_4, d_5, d_6, d_7, d_3, d_2),$$

whereas the unique extremal caterpillar with respect to the hyper-Wiener index is the caterpillar T_2 with backbone vertex degrees in the order

$$(d_{2,1}, \ldots, d_{2,7}) = (d_1, d_4, d_5, d_7, d_6, d_3, d_2).$$

Here $d_{i,j}$ is the degree of the j-th vertex on the backbone of T_i, $1 \leq j \leq 7$, $i = 1, 2$.

Similar to the Wiener index of caterpillars, $W_f(T)$ can also be directly computed from its degree sequence. We leave the technical details as an exercise to interested readers.

Lemma 2.6.1 *Let T be a caterpillar on n vertices and $v_1, \ldots, v_k$ the vertices on the backbone of T in this order. Then,*

$$W_f(T) = \frac{1}{2} \sum_{i=1}^{k} \sum_{j=1}^{k} f(|j-i|+2)(d_i-2)(d_j-2)$$

$$+ \sum_{i=1}^{k} \sum_{j=0}^{k+1} f(|j-i|+1)(d_i-2) \qquad (2.8)$$

$$+ \sum_{i=0}^{k} \sum_{j=i+1}^{k+1} f(j-i) - \frac{1}{2} f(2)(n-k-2)$$

for any function $f(x)$ with $d_\ell = \deg(v_\ell)$.

Making use of the expression (2.8), one can also establish the ∨-property for the extremal caterpillars, of a given degree sequence, with respect to $W_f(\cdot)$.

Theorem 2.6.1 *Let $x_1 \geq x_2 \geq \cdots \geq x_k \geq 0$ be integers with $k \geq 3$ and let $f(x)$ be a strictly increasing and convex function. Further, let S_k be the set of all permutations of $\{1, \ldots, k\}$ and suppose that $(y_1, \ldots, y_k)$ is a permutation of $(x_1, \ldots, x_k)$ with $y_1 \geq y_k$ and*

$$\frac{1}{2} \sum_{i=1}^{k} \sum_{j=1}^{k} f(|j-i|+2) y_i y_j + \sum_{i=1}^{k} \sum_{j=0}^{k+1} f(|j-i|+1) y_i$$

$$= \max_{\pi \in S_k} \left(\frac{1}{2} \sum_{i=1}^{k} \sum_{j=1}^{k} f(|j-i|+2) x_{\pi(i)} x_{\pi(j)} + \sum_{i=1}^{k} \sum_{j=0}^{k+1} f(|j-i|+1) x_{\pi(i)} \right).$$

Then there exists a $t \in \{2, 3, \ldots, k-1\}$ such that

$$y_1 \geq y_2 \geq \cdots \geq y_{t-1} \geq y_t \leq y_{t+1} \leq \cdots \leq y_k.$$

Moreover, if $k \geq 5$, then $t \neq k-1$.

Proof:

First, consider the permutation

$$(z_1, \ldots, z_k) = (y_1, \ldots, y_{\ell-1}, y_{\ell+1}, y_\ell, y_{\ell+2}, \ldots, y_k)$$

of $(x_1, \ldots, x_k)$.

As $(y_1, \ldots, y_k)$ achieves the maximum value, we must have

$$0 \le \frac{1}{2} \sum_{i=1}^{k} \sum_{j=1}^{k} f(|j-i|+2)(y_i y_j - z_i z_j) + \sum_{i=1}^{k} \sum_{j=0}^{k+1} f(|j-i|+1)(y_i - z_i)$$

$$= (y_{\ell+1} - y_\ell)\left(\sum_{i=1}^{\ell-1} (f(\ell-i+3) - f(\ell-i+2))y_i \right. \tag{2.9}$$

$$\left. - \sum_{i=\ell+2}^{k} (f(i-\ell+2) - f(i-\ell+1))y_i + f(\ell+2) - f(k-\ell+2) \right).$$

Letting

$$g(\ell) := \sum_{i=1}^{\ell-1} (f(\ell-i+3) - f(\ell-i+2))y_i - \sum_{i=\ell+2}^{k} (f(i-\ell+2) - f(i-\ell+1))y_i$$

$$+ f(\ell+2) - f(k-\ell+2)$$

for $1 \le \ell \le k-1$, it is easy to see that

$$g(1) < 0, \; g(k-1) > 0, \text{ and } g(\ell) < g(\ell+1).$$

Consequently, there exists a $t' \in \{2, 3, \ldots, k-2\}$ such that

$$g(t'-1) < 0, \qquad g(t'+1) > 0.$$

Together with (2.9) we obtain

$$\begin{aligned} y_{\ell+1} - y_\ell &\le 0 &\text{for } 1 \le \ell \le t'-1, \\ y_{\ell+1} - y_\ell &\ge 0 &\text{for } t'+1 \le \ell \le k-1, \end{aligned}$$

implying

$$y_1 \ge y_2 \ge \cdots \ge y_{t'-1} \ge y_{t'} \qquad \text{and} \qquad y_{t'+1} \le y_{t'+2} \le \cdots \le y_k.$$

The conclusion follows from the following two cases:

- If $y_{t'} \le y_{t'+1}$, then $t = t'$.

- If $y_{t'} \ge y_{t'+1}$, then $t = t' + 1$. Since $g(k-2) > 0$ for $y_1 \ge y_k$ and $k \ge 5$, we obtain that $t' \ne k-2$ in this case. Therefore, $2 \le t \le k-2$.

$\square$

Before concluding our discussion here, we point out that analogous statements can be easily established, through the same arguments, for decreasing and concave functions. We list two such statements here.

Proposition 2.6.3 *Let $f(x)$ be a strictly decreasing and concave function (i.e., the decrements $f(x) - f(x+1)$ are non-decreasing). If T is a tree that minimizes*

$$W_f(T) = \sum_{\{u,v\} \subseteq V(T)} f(d(u,v))$$

among all trees with degree sequence $(d_1, \ldots, d_n)$, then T must be a caterpillar.

Proposition 2.6.4 *Let $f(x)$ be a non-increasing and concave function (i.e., the decrements $f(x) - f(x+1)$ are non-decreasing). Among all trees with degree sequence $(d_1, \ldots, d_n)$ there always exists a caterpillar that achieves the minimum value of*

$$W_f(T) = \sum_{\{u,v\} \subseteq V(T)} f(d(u,v)).$$

2.6.2 Greedy trees

We now move on to the greedy trees. The main theorem relies on the interesting statement below, established in [94], showing that the greedy tree is indeed greedy with respect to the distances in the following sense.

Theorem 2.6.2 *Let $d_1 \geq d_2 \geq \cdots \geq d_n$ be positive integers such that $\sum_i d_i = 2(n-1)$, and let k be another arbitrary positive integer. Among all trees with degree sequence $(d_1, d_2, \ldots, d_n)$, the greedy tree has the largest number $p_k(T)$ of pairs (u,v) of vertices such that $d(u,v) \leq k$.*

Before considering the proof of Theorem 2.6.2, let us first introduce the following important consequence that is our main statement of this section.

Corollary 2.6.1 *Let $f(x)$ be any non-negative, non-decreasing function of x. Then, the index*

$$W_f(T) = \sum_{\{u,v\} \subseteq V(T)} f(d(u,v))$$

is minimized by the greedy tree among all trees with a given degree sequence.

Proof:

First, note that

$$W_f(T) = \sum_{k \geq 0} (f(k+1) - f(k)) \left| \{ \{u,v\} \subseteq V(T) : d(u,v) > k \} \right|.$$

Since $f(k+1) - f(k)$ is non-negative for all k (we set $f(0) = 0$) as f is non-decreasing, the conclusion follows from Theorem 2.6.2 by the definition of $p_k(T)$. $\qquad \square$

Just as in the previous section, the above corollary includes the classical Wiener index, the hyper-Wiener index, and the generalized Wiener index with $\alpha > 0$.

Remark 2.6.3 *Another interesting fact to note is that Corollary 2.6.1 is, as a matter of fact, equivalent to Theorem 2.6.2. To see this, let*

$$f_k(x) = \begin{cases} 0 & x \leq k, \\ 1 & x > k, \end{cases}$$

so that

$$W_{f_k}(T) = \binom{n}{2} - p_k(T)$$

for any tree T of order n. Hence p_k is maximized if $W_{f_k}(T)$ is minimized and vice versa.

Now, in order to establish the extremality of the greedy trees, we should try to understand the structural properties that define a greedy tree. From Definition 2.1.1, it is probably obvious that the following is true in a greedy tree:

Proposition 2.6.5 *The vertex degrees along any maximal path (of a greedy tree) are increasing from one end to the middle and then decreasing. Furthermore, vertices closer to the middle (regardless of which side they are on) have larger degrees. That is:*

- *for any path $P = u_p, u_{p-1}, \ldots, u_1, v_1, \ldots, v_p$ of odd length $(2p-1)$, we must have (possibly after reversing the order of the vertices)*

$$\deg(u_1) \geq \deg(v_1) \geq \deg(u_2) \geq \deg(v_2) \geq \cdots > \deg(u_p) = \deg(v_p) = 1;$$

- *for any path $P = u_{p+1}, u_p, \ldots, u_1, v_1, \ldots, v_p$ of even length $(2p)$, we must have (possibly after reversing the order of the vertices)*

$$\deg(u_1) \geq \deg(v_1) \geq \deg(u_2) \geq \deg(v_2) \geq \cdots > \deg(v_p) = \deg(u_{p+1}) = 1.$$

To see this, one simply recalls that in a greedy tree with its designated root, any maximal path has its middle point at the smallest height (among vertices on this path) and the vertex degrees decrease from this middle point to both ends of the path. We skip the details. It was shown in [110] that this rather obvious necessary condition is also sufficient for a tree, with a given degree sequence, to be greedy.

Next, we will introduce and prove a sufficient and necessary condition for a tree (with a given degree sequence) to be greedy in terms of the level-greedy trees. There are a number of different proofs in the literature. We briefly introduce the idea of the most elementary one.

Theorem 2.6.3 *Every greedy tree with a given degree sequence is also level-greedy with any choice of root or edge-root.*

On the other hand, if a tree T with a given degree sequence is level-greedy with respect to any possible choice of root or edge-root, then T is a greedy tree.

Proof:

The first part is obvious from the definitions of greedy and level-greedy trees. We leave the proof as an exercise.

To see the other direction, let us start with a tree T with a given degree sequence that is level-greedy with respect to any root or edge-root. Following the definition of a greedy tree, we first choose the root r to be:

- one of the vertices with the largest degree;

- among the vertices with the largest degree, one with the largest sum of degrees of its neighbors;

- among the vertices with the largest degree and largest sum of neighboring degrees, one with the largest sum of degrees of vertices at distance 2;

- etc.

Note that T is level-greedy with root r, so to show that T is greedy with respect to root r we only need to establish the fact that vertices of smaller height have greater degrees. Suppose, for contradiction, that this is not the case. Then we have two vertices u and v where u is of smaller height and $\deg(u) < \deg(v)$. Without loss of generality we may assume the following:

1. v has height exactly 1 more than u;

2. u has the smallest degree among vertices of the same height and v has the largest degree among vertices of the same height;

3. for all vertices of smaller height than u and v we have the fact that "vertices of smaller height have greater degrees".

From our assumption, the path $P(u,v)$ is of odd length. Let $e = st$ be the edge in the middle of $P(u,v)$. We now consider T', the same tree as T but edge-rooted at e. Further, assume that, in T', u is a descendant of s and v is a descendant of t. Also note that u and v are of the same height in T'. By assumption (3) above we know $\deg(s) \geq \deg(t)$. Consider two cases:

- If $\deg(s) > \deg(t)$, then since T' is level-greedy, and u (as a descendant of s) and v (as a descendant of t) are of the same height, we must have $\deg(u) \geq \deg(v)$, a contradiction;

- If $\deg(s) = \deg(t)$, consider now the sum of degrees of descendants of s and t, respectively, of each height. Again by assumption (3) above we know the sum on the "s side" is always at least as large as that on the "t side":

- if strict inequality holds for descendants of any height, then the same reasoning as in the previous case yields a contradiction;

- otherwise, we have the same number of descendants of s and t, of each height, and they are all of the same degree. Now for descendants of the same height as u and v, since $\deg(u) < \deg(v)$ and T' is level-greedy, we must have degree sums of vertices of any given height in T', that are larger on the "t side" than on the "s side" with at least one strict inequality. Note that s is closer to r than t is in T, so this contradicts our choice of r in the first place.

$\square$

With Theorem 2.6.3, we can now prove Theorem 2.6.2 by establishing that the extremal trees (with a given degree sequence, that maximize p_k) must be level-greedy. This was first done in [94].

Lemma 2.6.2 *Among all rooted trees whose outdegrees at each level i are given by a multiset $\{a_{i1}, a_{i2}, \ldots, a_{i\ell_i}\}$ as in Definition 2.1.2, the level-greedy tree maximizes the value of $p_k(T)$.*

First, we prove a technical lemma that was also introduced in [94].

Lemma 2.6.3 *Suppose that the sequences $(x_1, x_2, \ldots, x_m)$, $(y_1, y_2, \ldots, y_m)$, $(x'_1, x'_2, \ldots, x'_m)$ and $(y'_1, y'_2, \ldots, y'_m)$ of non-negative real numbers satisfy*

$$\sum_{j=1}^{h} x_j \geq \sum_{j=1}^{h} x'_{\upsilon(j)} \qquad and \qquad \sum_{j=1}^{h} y_j > \sum_{j=1}^{h} y'_{\sigma(j)} \qquad (2.10)$$

for all $1 \leq h \leq m$ and all permutations σ of $\{1, 2, \ldots, m\}$. Then

$$x_1 y_1 + x_2 y_2 + \cdots + x_m y_m \geq x'_1 y'_1 + x'_2 y'_2 + \cdots + x'_m y'_m. \qquad (2.11)$$

Proof:

Suppose that $x'_1, x'_2, \ldots, x'_m$ and $y'_1, y'_2, \ldots, y'_m$ are such that the sum

$$x'_1 y'_1 + x'_2 y'_2 + \cdots + x'_m y'_m \qquad (2.12)$$

attains its maximum under the stated restrictions. We can assume, without loss of generality, that

$$x'_1 \geq x'_2 \geq \ldots \geq x'_m \quad and \quad y'_1 \geq y'_2 \geq \ldots \geq y'_m$$

by the rearrangement inequality, which states that the maximum of (2.12) under permutations of the x'_i and y'_j is attained when both are ordered in the same way (e.g., both increasing or both decreasing).

If $(x_1, \ldots, x_m) \neq (x'_1, \ldots, x'_m)$, let h be the smallest index such that

$$x_1 + x_2 + \ldots + x_h > x'_1 + x'_2 + \ldots + x'_h,$$

and let $\epsilon > 0$ be the difference between the two sides of the inequality. Replacing x'_h by $x'_h + \epsilon$ and x'_{h+1} by $x'_{h+1} - \epsilon$, we obtain a new $(2m)$-tuple of numbers, still satisfying the requirements, while the sum

$$x'_1 y'_1 + x'_2 y'_2 + \cdots + x'_m y'_m$$

changes by $\epsilon(y'_h - y'_{h+1}) \geq 0$. This process can be repeated until we have

$$x_1 = x'_1, x_2 = x'_2, \ldots, x_m = x'_m, y_1 = y'_1, y_2 = y'_2, \ldots, y_m = y'_m.$$

$\square$

We are now ready to prove Lemma 2.6.2.

Proof of Lemma 2.6.2:

We proceed by considering the number of paths between vertices at different levels i and j that are of length at most k.

Given i and j, if $i + j \leq k$, then $d(u, v) \leq k$ for any u at level i and v at level j.

Otherwise, a vertex u at level i and a vertex v at level j satisfy $d(u, v) \leq k$ if and only if they have the same ancestor at level $\lceil (i + j - k)/2 \rceil$. We will count the number of such pairs of vertices below.

For this purpose, let $w_1, w_2, \ldots, w_m$ be the vertices at level r, and denote by $x_1, x_2, \ldots, x_m$ and $y_1, y_2, \ldots, y_m$ the number of their respective successors at level i and level j. Then, the number of pairs we have to count is

$$x_1 y_1 + x_2 y_2 + \cdots + x_m y_m$$

if $i \neq j$, and otherwise

$$\binom{x_1}{2} + \binom{x_2}{2} + \cdots + \binom{x_m}{2}.$$

However, in the latter case, as the sum $x_1 + x_2 + \cdots + x_m$ is constant under reshuffling (i.e., the process of changing the tree while keeping the same root and level-degree sequence), maximizing this sum is equivalent to maximizing

$$x_1^2 + x_2^2 + \cdots + x_m^2 = x_1 y_1 + x_2 y_2 + \cdots + x_m y_m,$$

so the case $i = j$ can be treated in the same way as the $i \neq j$ case.

Under all possible "reshuffled" trees, it is clear that the level-greedy tree maximizes $x_1 + x_2 + \cdots + x_h$ and $y_1 + y_2 + \cdots + y_h$ for all $1 \leq h \leq m$. Hence, the result will follow as a consequence of Lemma 2.6.3. As it is easy to see that (2.10) is satisfied and hence (2.11) implies that the level-greedy tree indeed maximizes the number of pairs of vertices at levels i and j that have a common ancestor at level $r = \lceil (i + j - k)/2 \rceil$, for every i and j.

That is, p_k is maximized by the level-greedy tree. $\qquad\qquad\qquad\square$

The very same arguments will justify the following similar statements for edge-rooted level-greedy trees. We leave the details as an exercise.

Lemma 2.6.4 *Among all edge-rooted trees whose outdegrees at each level are given by a multiset $\{a_{i1}, a_{i2}, \ldots, a_{i\ell_i}\}$ as in Definition 2.1.2, the greedy tree maximizes the value of $p_k(T)$.*

We conclude this section by noting that Theorem 2.6.2 is a direct consequence of Lemma 2.6.2, Lemma 2.6.4, and Theorem 2.6.3. Of course, completely analogous results hold for extremal trees of a given degree sequence that maximize $W_f(\cdot)$ for a non-increasing function f, such as in the case of the Harary index. The proof is left as an exercise.

Theorem 2.6.4 *Let $f(x)$ be any non-negative, non-increasing function of x. Then, the index*

$$W_f(T) = \sum_{\{u,v\} \subseteq V(T)} f(d(u,v))$$

is maximized by the greedy tree among all trees with a given degree sequence.

2.6.3 Comparing greedy trees of different degree sequences and applications

So far the introduction of $W_f(\cdot)$ is probably the most general concept we have seen in terms of extremal problems with respect to distance-based topological indices. Another direction of general approaches is to obtain generalized characterization of extremal structures under various constraints. This is done through comparing the greedy trees of different degree sequences.

Recall the partial ordering between degree sequences (on the same number of vertices) introduced as majorization. In the rest of this section, we will see that they also define a partial ordering on the extremal values of distance-based indices among trees of different degree sequences and this partial ordering can be used to identify many other extremal structures.

We start with a comparison between greedy trees of different degree sequences, with respect to $p_k(T)$.

Theorem 2.6.5 *Given two different degree sequences π and π' with $\pi \lhd \pi'$, we have*

$$p_k(T_\pi^*) \leq p_k(T_{\pi'}^*)$$

where T_π^* and $T_{\pi'}^*$ are the greedy trees with degree sequences π and π', respectively.

Proof:

By Lemma 1.4.1, it is sufficient to show the statement for two degree sequences

$$\pi = (d_0, \ldots d_{n-1}) \lhd (d'_0, \ldots, d'_{n-1}) = \pi'$$

such that they differ only at the j-th and k-th entries with $d'_j = d_j + 1$, $d'_k = d_k - 1$ for some $j < k$.

Let $T_{\pi'}$ be the tree obtained from T^*_π by removing the edge vw and adding an edge uw, where u and v are the vertices corresponding to d_j and d_k, respectively, and w is a child of v (Figure 2.17).

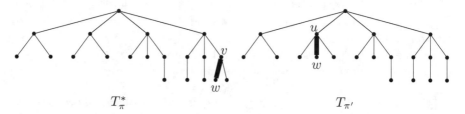

FIGURE 2.17
The trees T^*_π and $T_{\pi'}$ with $\pi = (4,4,4,3,3,3,2,2,2,1,\ldots,1)$ and $\pi' = (4,4,4,4,3,2,2,2,2,1,\ldots,1)$.

Note that $T_{\pi'}$ has degree sequence π', but it is not necessarily a greedy tree.

Let T' be the tree obtained from T^*_π after removing w and its descendants. Then, it is an easy exercise, based on the structure of the greedy tree T^*_π, to check that

$$p_k(T', u) \geq p_k(T', v) \text{ for all } k \geq 1,$$

where $p_k(T, x)$ is the number of vertices in T at distance $\leq k$ from x. Consequently, we have

$$p_k(T_{\pi'}) \geq p_k(T^*_\pi)$$

following straightforward computation. Intuitively, this is because $T_{\pi'}$ places w and its descendants closer to more vertices at shorter distance than T^*_π does, and consequently generates more short paths.

On the other hand, by the extremality of the greedy tree $T^*_{\pi'}$ with respect to $p_k(\cdot)$ we now have

$$p_k(T^*_{\pi'}) \geq p_k(T_{\pi'}) \geq p_k(T^*_\pi).$$

$\square$

As a first example of the applications of this powerful result, consider all possible degree sequences of trees of order n and maximum degree Δ. It is easy to see that the degree sequence $(\Delta, \Delta, \ldots, \Delta, m, 1, \ldots, 1)$ ($m \in \{1, 2, \ldots, \Delta - 1\}$ chosen to be congruent to $n - 1$ modulo $\Delta - 1$) majorizes all other degree sequences. Hence, Theorem 2.6.6 immediately implies the following.

Corollary 2.6.2 *The "complete Δ-ary tree" maximizes $p_k(T)$ among trees with maximum degree Δ.*

Here the *complete Δ-ary tree* with a given maximum degree Δ (also called the *good tree* or *Volkmann tree*) is defined in a similar way as the greedy tree, except that the vertices $v, v_1, \ldots$ take the maximum degree Δ until there are not enough vertices (Figure 2.18). As a result, the complete Δ-ary tree has degree sequence $(\Delta, \Delta, \ldots, \Delta, m, 1, \ldots, 1)$ for some $m \in \{1, 2, \ldots, \Delta - 1\}$.

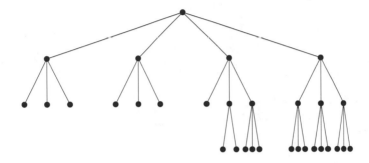

FIGURE 2.18
A complete 4-ary tree.

Similar to the previous section, the extremal statements with respect to $p_k(\cdot)$ immediately transform into statements for distance-based indices.

Corollary 2.6.3 *Let $f(x)$ be any non-negative, non-increasing (non-decreasing) function of x. Then the graph invariant*

$$W_f(T) = \sum_{\{u,v\} \subseteq V(T)} f(d(u,v))$$

is maximized (minimized) by the complete Δ-ary tree among all trees with given order and maximum degree Δ.

Similar to the above discussion, we may immediately transform Theorem 2.6.5 into the following version in terms of $W_f(\cdot)$.

Theorem 2.6.6 *Given two different degree sequences π and π' with $\pi \lhd \pi'$, then:*

• *for a non-negative and non-increasing function $f(x)$ we have*

$$W_f(T_\pi^*) \le W_f(T_{\pi'}^*);$$

and

- *for a non-negative and non-decreasing function $f(x)$ we have*

$$W_f(T_\pi^*) \geq W_f(T_{\pi'}^*).$$

Here, T_π^* and $T_{\pi'}^*$ are the greedy trees with degree sequences π and π', respectively.

It is easy to see from Corollary 2.6.2 that among different classes of trees, to apply Theorem 2.6.6 one only needs to identify the extremal degree sequence, under the given constraints, that majorizes all other degree sequences. We show a few more well-known applications along this line. A very brief justification is provided for completeness.

Corollary 2.6.4 *Let $f(x)$ be any non-negative and non-increasing (non-decreasing) function of x. Then,*

$$W_f(T) = \sum_{\{u,v\} \subseteq V(T)} f(d(u,v))$$

is maximized (minimized) by the star $K_{1,n-1}$ among all trees of a given order.

Proof:

For every tree of order n, its degree sequence π is majorized by $\pi' = (n - 1, 1, \ldots, 1)$ with "equality" if and only if the tree is the star $K_{1,n-1}$.   □

It has been of interest to study extremal problems among structures with a given number of leaves, a given independence number, or a given matching number. For this purpose we let $\mathcal{T}_{n,s}^{(1)}$ be the set of all trees of order n with s leaves, $\mathcal{T}_{n,\alpha}^{(2)}$ be the set of all trees of order n with independence number α and $\mathcal{T}_{n,\beta}^{(3)}$ be the set of all trees of order n with matching number β.

Similar to Corollary 2.6.4, useful consequences follow from Theorem 2.6.6.

Corollary 2.6.5 *Let $f(x)$ be any non-negative, non-increasing (non-decreasing) function of x. Then,*

$$W_f(T) = \sum_{\{u,v\} \subseteq V(T)} f(d(u,v))$$

is maximized (minimized) by the tree T_s^* in $\mathcal{T}_{n,s}^{(1)}$, where T_s^* is the starlike tree obtained from t paths of order $q+2$ and $s-t$ paths of order $q+1$ by identifying one end of each of the s paths. Here $n - 1 = sq + t, 0 \leq t < s$.

Proof:

Let T be any tree in $\mathcal{T}_{n,s}^{(1)}$ with degree sequence $\pi_1 = (d_0, \ldots, d_{n-1})$. Thus,

$$d_{n-s-1} > 1 \text{ and } d_{n-s} = \cdots = d_{n-1} = 1.$$

Let T_π^* be a greedy tree with degree sequence $\pi = (s, 2, \ldots, 2, 1, \ldots, 1)$, where the number of 1s in π is s. It is easy to see that $\pi_1 \triangleleft \pi$. The conclusion then follows from Theorem 2.6.6. □

Corollary 2.6.6 *Let* $f(x)$ *be any non-negative, non-increasing (non-decreasing) function of* x. *Then*

$$W_f(T) = \sum_{\{u,v\} \subseteq V(T)} f(d(u,v))$$

is maximized (minimized) by the tree T_α^* *in* $\mathcal{T}_{n,\alpha}^{(2)}$, *where* T_α^* *is* T_π^* *with degree sequence* $\pi = (\alpha, 2, \ldots, 2, 1, \ldots, 1)$ *with numbers* $n - \alpha - 1$ *of* $2's$ *and* α *of* $1's$, *i.e.,* T_π^* *is obtained from the star* $K_{1,\alpha}$ *by adding* $n - \alpha - 1$ *pendant edges to* $n - \alpha - 1$ *leaves of* $K_{1,\alpha}$.

Proof:

For any tree T of order n with independence number α, let I be an independent set of T with size α and $\tau = (d_0, \ldots, d_{n-1})$ be the degree sequence of T. If there exists a leaf u with $u \notin I$, then there exists a vertex $v \in I$ with $(u, v) \in E(T)$. Hence $I \cup \{u\} \setminus \{v\}$ is an independent set of T with size α. Repeating this argument, one can always construct an independent set of T with size α that contains all leaves of T. Hence, there are at most α leaves. The conclusion then follows from similar arguments as that of Corollary 2.6.5. □

Corollary 2.6.7 *Let* $f(x)$ *be any non-negative, non-increasing (non-decreasing) function of* x. *Then*

$$W_f(T) = \sum_{\{u,v\} \subseteq V(T)} f(d(u,v))$$

is maximized (minimized) by the tree T_β^* *in* $\mathcal{T}_{n,\beta}^{(3)}$, *where* T_β^* *is the greedy tree* T_π^* *with degree sequence* $\pi = (n - \beta, 2, \ldots, 2, 1, \ldots, 1)$. *Here, the number of 1s in* π *is* $n - \beta$. *That is,* T_π^* *is obtained from the star* $K_{1,n-\beta}$ *by adding* $\beta - 1$ *pendant edges to* $\beta - 1$ *leaves of* $K_{1,n-\beta}$.

Proof:

For any tree T of order n with matching number β, let $\tau = (d_0, \ldots, d_{n-1})$ be the degree sequence of T. Let M be a matching of T with size β. Since T is connected, there are at least β vertices in T such that their degrees are at least 2, as each matching edge of M has to contain at least one such vertex. Hence, $d_{\beta-1} \geq 2$ and $\tau \triangleleft \pi$. The conclusion follows from Theorem 2.6.6. □

Let us remark that the previous two corollaries are in fact equivalent: for a tree T with n vertices, the independence number $\alpha(T)$ and the matching number $\beta(T)$ satisfy (see Exercise 4 in Chapter 1)

$$\alpha(T) + \beta(T) = n,$$

hence $T_{n,\alpha}^{(2)}$ and $T_{n,n-\alpha}^{(3)}$ coincide.

Theorem 2.6.6 and the general approach discussed above can be employed to deal with extremal questions in many other classes of trees, some of which we list below. It is a good exercise to identify the degree sequence under the different constraints and their corresponding extremal structures.

- Among trees with a given number of branching vertices;

- Among trees with all vertex degrees odd;

- Among trees with a given number of vertices with even degrees;

- Among trees with a given minimum degree for internal vertices;

- Among trees with a given number of segments.

2.7   The inverse problem

There are many other problems in chemical graph theory, in addition to the extremal problems that we have discussed so far, that are worth exploring. One of them is the so-called inverse Wiener index problem [46]:

Given a positive integer n, can we find a structure (graph) with Wiener index n?

One of the first related observations states that except for 49 of them, all positive integers can be represented as the Wiener index of some tree. The 49 integers that are not Wiener index of any tree are

2, 3, 5, 6, 7, 8, 11, 12, 13, 14, 15, 17, 19, 21, 22, 23, 24, 26, 27, 30, 33, 34, 37, 38, 39, 41, 43, 45, 47, 51, 53, 55, 60, 61, 69, 73, 77, 78, 83, 85, 87, 89, 91, 99, 101, 106, 113, 147, 159.

This was independently proved in [103] and [115] through completely different approaches. Based on those studies, it was pointed out that the molecular graphs of most practical interest have natural restrictions on their degrees (i.e., trees with limited maximal degree) or have hexagonal or pentagonal cycles. Mathematically, this inspired the study of the inverse Wiener index problem for two types of structures in [107]: trees with vertex degrees ≤ 3 (Figure 2.19), and a type of graphs involving hexagonal chains (the interested reader is referred to [107] for details of the latter).

In this section, we show a solution to the inverse problem by introducing a specific type of trees, representing their Wiener index mathematically, and eventually showing that the resulting formula represents almost all positive integers.

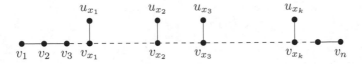

FIGURE 2.19
Caterpillar tree with degree ≤ 3.

Consider the family of trees $T = T(n, x_1, x_2, \ldots, x_k)$, where

$$V = \{v_1, \ldots, v_n\} \cup \{u_{x_1}, \ldots, u_{x_k}\},$$

$$E = \{v_i v_{i+1}, 1 \leq i \leq n-1\} \cup \{v_{x_i} u_{x_i}, 1 \leq i \leq k\},$$

where n and $x_i, 1 \leq i \leq k$, are integers such that $1 < x_1 < \ldots < x_k < n$ (Figure 2.19).

The following theorem was established in [107]. We will briefly discuss the ideas of the arguments in this section.

Theorem 2.7.1 *Every sufficiently large integer n is the Wiener index of a caterpillar tree with degree ≤ 3.*

First, we explicitly represent the Wiener index of these trees: for $T = T(n, x_1, x_2, \ldots, x_k)$, we have

$$W(T) = \sum_{1 \leq i \leq j \leq n} d(v_i, v_j) + \sum_{i=1}^{n} \sum_{j=1}^{k} d(v_i, u_{x_j}) + \sum_{1 \leq i \leq j \leq k} d(u_{x_i}, u_{x_j})$$

$$= \frac{n^3 - n}{6} + \sum_{i=1}^{n} \sum_{j=1}^{k} (1 + |x_j - i|) + \sum_{1 \leq i < j \leq k} (2 + x_j - x_i),$$

which simplifies to

$$\frac{n^3}{6} + \frac{kn^2}{4} + \frac{(6k-1)n}{6} - \frac{k^3 - 12k^2 + 14k}{12} + \sum_{j=1}^{k} \left(x_j + j - 1 - \frac{k+n}{2} \right)^2. \quad (2.13)$$

By taking $k = 8$ and $n = 2s$ in (2.13) and setting $y_j := x_j + j - 5 - s$, we have

$$W(T(n, x_1, \ldots, x_8)) = \frac{4s^3}{3} + 8s^2 + \frac{47s}{3} + 12 + \sum_{j=1}^{8} y_j^2 \quad (2.14)$$

under the conditions that

$$-3 - s \leq y_1 < y_2 < \ldots < y_8 \leq 3 + s$$

and without any two consecutive y_j (since no two x_j's may be the same).

We now introduce the following modification of Lagrange's famous four-square theorem, shown in [107]. We skip the number-theoretical proof here.

Lemma 2.7.1 *Let $N > 103$ and $4 \nmid N$. Then N can be written as $a_1^2 + a_2^2 + a_3^2 + a_4^2$ with non-negative integers $a_1 < a_2 < a_3 < a_4$ and $a_2 \geq 2$.*

By setting $z_1 = -a_3$, $z_2 = -a_1$, $z_3 = a_2$ and $z_4 = a_4$ with a_1, a_2, a_3, a_4 satisfying the conditions in the above lemma, we immediately have the following.

Corollary 2.7.1 *If $4 \nmid N$, $N > 103$, one can always find integers z_1, z_2, z_3, z_4 such that $N = z_1^2 + \ldots + z_4^2$, $z_1 < \ldots < z_4$ and no two of the z_i are consecutive.*

With these preparations, one can show that $\sum_{j=1}^8 y_j^2$ can represent all numbers in infinitely many intervals of positive integers.

Proposition 2.7.1 *Let $K \geq 15$. Then every integer N in the interval*

$$[4K^2 - 8K + 112, 5K^2 - 16K + 21]$$

can be written as $y_1^2 + \ldots y_8^2$, where the y_i are integers satisfying

$$-K \leq y_1 < y_2 < \ldots < y_8 \leq K$$

and no two of them are consecutive.

Proof:

Take $y_1 = -K$, $y_7 = K - 2$, $y_8 = K$ and either $y_2 = -K + 2$ or $y_2 = -K + 3$. By Corollary 2.7.1 and the additional observation that $z_4 \leq \lfloor \sqrt{N} \rfloor$ and $|z_1| \leq \lfloor \sqrt{N} \rfloor - 1$, one finds that every integer $M \in [104, (K-3)^2 - 1]$, $4 \nmid M$, can be written as $y_3^2 + \ldots + y_6^2$, where

$$-K = y_1 < y_2 < -K + 4 < y_3 < y_4 < y_5 < y_6 < K - 3 < y_7 < y_8 = K$$

(no two of them being consecutive). Now

$$(-K)^2 + (-K+2)^2 + (K-2)^2 + K^2 = 4K^2 - 8K + 8 \equiv 0 \mod 4$$

and

$$(-K)^2 + (-K+3)^2 + (K-2)^2 + K^2 = 4K^2 - 10K + 13 \equiv 2K + 1 \mod 4.$$

So all integers $\not\equiv 0 \mod 4$ in the interval $[4K^2 - 8K + 112, 5K^2 - 14K + 16]$ and all integers $\not\equiv 2K + 1 \mod 4$ in the interval $[4K^2 - 10K + 117, 5K^2 - 16K + 21]$ can be written in the required way. Since $0 \not\equiv 2K + 1 \mod 4$, this means that in fact all integers in the interval $[4K^2 - 8K + 112, 5K^2 - 16K + 21]$ can be written in the required way, which proves the claim. $\qquad \square$

For any large enough integer, we may show that it has to fall into one of the above intervals, consequently leading to the following.

Theorem 2.7.2 *All integers ≥ 3856 are Wiener indices of trees of the form $T(n, x_1, \ldots, x_8)$ $(x_1 < x_2 < \ldots < x_8)$ and thus Wiener indices of chemical trees.*

Proof:

By the preceding proposition, any integer in the interval $[4K^2 - 8K + 112, 5K^2 - 16K + 21]$ can be written as $y_1^2 + \ldots + y_8^2$, where the y_i satisfy our requirements and $-K \leq y_1 < \ldots < y_8 \leq K$. If we take the union of these intervals over $21 \leq K \leq s + 3$, we see that in fact any integer in the interval $[1708, 5s^2 + 14s + 18]$ can be written as $y_1^2 + \ldots y_8^2$, where the y_i satisfy our requirements and $-3 - s \leq y_1 < \ldots < y_8 \leq s + 3$. Short computer calculations show that, for $s \geq 7$, even any integer in the interval $[224, 5s^2 + 14s + 18]$ can always be written that way. But this means that for any $s \geq 7$, all integers in the interval

$$\left[\frac{4s^3}{3} + 8s^2 + \frac{47s}{3} + 236, \frac{4s^3}{3} + 13s^2 + \frac{89s}{3} + 30 \right]$$

are Wiener indices of trees of the form $T(n, x_1, \ldots, x_8)$. Taking the union over all these intervals, we see that all integers ≥ 12567 are contained in an interval of that type. By an additional computer search ($n \leq 40$ will do) in the remaining interval, one can get this number down to 3856. $\square$

Remark 2.7.1 *By checking $k = 4, 5, 6, 7$ and finally all $n \leq 17$, one obtains a list of 250 integers (the largest being 927) that are not Wiener indices of trees of the form $T(n, x_1, \ldots, x_k)$ with maximal degree ≤ 3. Further computer search gives a list of 127 integers that are not Wiener indices of trees with maximal degree ≤ 3 – these are 16, 25, 28, 36, 40, 42, 44, 49, 54, 57, 58, 59, 62, 63, 64, 66, 80, 81, 82, 86, 88, 93, 95, 97, 103, 105, 107, 109, 111, 112, 115, 116, 118, 119, 126, 132, 139, 140, 144, 148, 152, 155, 157, 161, 163, 167, 169, 171, 173, 175, 177, 179, 181, 183, 185, 187, 189, 191, 199, 227, 239, 251, 255, 257, 259, 263, 267, 269, 271, 273, 275, 279, 281, 283, 287, 289, 291, 405 and the 49 values that cannot be represented as the Wiener index of any tree. This list reduces to the following values if one considers also trees with maximal degree $= 4$: 25, 36, 40, 49, 54, 57, 59, 80, 81, 93, 95, 97, 103, 105, 107, 109, 132, 155, 157, 161, 163, 167, 169, 171, 173, 177, 239, 251, 255 and 257.*

Exercises

1. Prove that
$$W(G) \geq n(n-1) - m$$
   holds for every connected graph G with n vertices and m edges.

2. Show that $W(T_1) > W(T_2)$ and $\Gamma(T_1) < \Gamma(T_2)$ for T_1 and T_2 in Figure 2.6.

3. Prove (2.1) for two k-ary trees T and T'.

4. Compute the sum of eccentricities of a path on n vertices.

5. Prove: the Wiener index of a tree with an odd number of vertices is always even.

6. Let $\mathcal{L}(T)$ be the line graph of a tree T, whose vertices are the edges of T, with an edge between two vertices of $\mathcal{L}(T)$ if and only if the corresponding edges are adjacent in T. Prove that the following formula holds for every tree with n vertices:

$$W(\mathcal{L}(T)) = W(T) - \binom{n}{2}.$$

7. Prove the following identity for the hyper-Wiener index of a tree T: for two vertices u and v, let $N(u,v)$ be the number of vertices (including u itself) for which the unique path to v passes through u. We have

$$WW(T) = \frac{1}{2} \sum_{\{u,v\} \subseteq V(T)} \left(d(u,v) + d(u,v)^2 \right)$$

$$= \sum_{\{u,v\} \subseteq V(T)} N(u,v)N(v,u).$$

8. Prove (2.3) in Section 2.4.

9. Following the discussion at the end of Section 2.4, characterize the chemical tree, with a given degree sequence, that maximizes the Wiener index.

10. Finish the proofs for $r_1 \leq \cdots \leq r_j$ in Theorem 2.5.4 and $s_1 \geq \cdots \geq s_{j'}$ in Theorem 2.5.5.

11. Prove Proposition 2.6.2.

12. Prove Lemma 2.6.1.

13. Prove Proposition 2.6.3.

14. Prove Proposition 2.6.4.

15. Prove Propostion 2.6.5.

16. Prove Lemma 2.6.4.

17. Prove Theorem 2.6.4.

18. Among degree sequences of each of the following classes of trees of given order, find the degree sequence that majorizes all others:

- Among trees with a given number of branching vertices;
- Among trees with all vertex degrees odd;
- Among trees with a given number of vertices with even degrees;
- Among trees with a given minimum degree for internal vertices;
- Among trees with a given number of segments.

3

Vertex degrees and the Randić index

3.1 Introduction

The best known degree-based index is probably the *Randić index*

$$R(G) = \sum_{uv \in E(G)} (\deg(u)\deg(v))^{-\frac{1}{2}},$$

introduced by Randić in 1975 [90]. Its more general version, known earlier as the branching index or connectivity index, is now called the *generalized Randić index*

$$R_\alpha(G) = \sum_{uv \in E(G)} (\deg_G(u)\deg_G(v))^{\alpha},$$

where $\alpha \neq 0$ can assume values other than $-\frac{1}{2}$.

With different values of α the concept $R_\alpha(\cdot)$ has appeared in various different instances. For instance, $R_1(G)$ is also known as the *second Zagreb index*. (The *first Zagreb index* is just the sum of degree squares; more general versions of it allow any exponent instead of 2.) For a tree T, $R_1(T)$ has also been known as the *weight* of a tree.

The Randić index gained further popularity following the work of Bollobás and Erdős [8,9]. It was shown that among graphs on n vertices with minimum degree at least 1, the star minimizes $R_{-1/2}$. As a generalization, in [67] the problem of minimizing $R_{-1/2}$ among graphs on n vertices with minimum degree at least k was discussed but not completely solved.

One of the reasons that the extremal problems with respect to the Randić index tend to be difficult may be the fact that, unlike what we have seen in many other chemical indices in this book, adding an edge to a graph does not necessarily increase nor decrease the Randić index. This "property" also leads to the disproofs of quite a number of published conjectures. Of course, in the general case of $R_\alpha(\cdot)$ with $\alpha \geq 0$, adding an edge does indeed increase the value of $R_\alpha(\cdot)$. This can be seen by noting that the addition of an edge can only increase the degrees of vertices which increase the value of some terms in the summation, and more edges means more (positive) terms in the summation. As an immediate consequence, the following is true. We leave the proof as an exercise.

Proposition 3.1.1 *For any $\alpha > 0$, among connected graphs on n vertices $R_\alpha(\cdot)$ is maximized by the complete graph and minimized by a tree.*

The cases with negative α are much more complicated for the obvious reason. As is the case for distance-based indices, there seem to be significantly more results on trees than on general graphs. Taking the weight $R_1(\cdot)$ as an example, paths and star continue to be extremal among general trees of a given order.

Proposition 3.1.2 *Among trees on n vertices:*

- *the path minimizes $R_1(\cdot)$;*

- *the star maximizes $R_1(\cdot)$.*

Proof:

- For the first part, suppose for contradiction that T is a tree on n vertices that minimizes the weight and that T is not a path.

  Then, similar to the proof of Proposition 2.3.3, let $P(u, v)$ be a longest path in T with leaves u and v, and let the vertices on this path be $u = v_0, v_1, v_2, \ldots, v_k, v_{k+1} = v$. Since T is not a path, there exists a vertex on $P(u, v)$, say v_i with the smallest subscript i such that v_i is of degree at least 3. It is obvious that i is at least 1 and at most k.

  Let w be a neighbor of v_i that is not on $P(u, v)$ and consider the tree $T' = T - v_i w + uw$ (Figure 3.1).

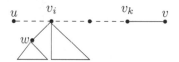

FIGURE 3.1
The path $P(u, v)$, v_i, and w.

From T to T' we only need to consider the edges xy whose corresponding term $\deg(x)\deg(y)$ changed. These are all the edges incident with the vertices u and v_i as the degree of u changed from 1 to 2 and the degree of v_i decreased by 1. Then, if $i > 1$, we have

$$R_1(T') - R_1(T)$$
$$= \deg_T(v_1) + \deg_T(w)(2 - \deg_T(v_i)) - \sum_{x \in N_T(v_i) \setminus \{w\}} \deg_T(x) < 0$$

as $\deg_T(v_i) \geq 3$ and $\deg_T(v_1) = 2$ from our constructions. This is a contradiction. The case for $i = 1$ is similar and we leave it as an exercise.

- For the second part, suppose for contradiction that T maximizes $R_1(\cdot)$ and T is not a star.

  Hence, T has at least two internal vertices. Consider now an internal vertex u of T, all but one of whose neighbors are leaves. Let the unique non-leaf neighbor of u be v and consider the tree T' obtained from T by removing all pendant edges at u and reattaching them at v (Figure 3.2).

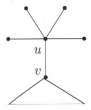

FIGURE 3.2
The tree T with internal vertices u and v.

Again, from T to T' we only need to consider the contribution to $R_1(\cdot)$ from edges incident with at least one of u and v. Then, by examining these edges and the corresponding product of adjacent degrees (the details are left as exercises), we have

$$R_1(T') - R_1(T) = (\deg_{T'}(v) - 1)(\deg_T(u) - 1) + (\deg_T(v) + \deg_T(u) - 1)$$
$$- \deg_T(v)\deg_T(u) + (\deg_T(u) - 1) \sum_{x \in N_{T'}(v)\setminus\{u\}} \deg_T(x)$$
$$= (\deg_T(u) - 1) \sum_{x \in N_T(v)\setminus\{u\}} \deg_T(x) > 0$$

as $\deg_T(u), \deg_T(v) \geq 2$. This is a contradiction.

□

As one would expect, to prove similar conclusions for the original Randić index, $R_{-1/2}(\cdot)$, or the general version, $R_\alpha(\cdot)$, will involve more complicated analysis and computation. Nevertheless, Proposition 3.1.2 and its proof provide a perfect example of the type of arguments in extremal problems with respect to distance-based indices.

Among trees of given order, it is known that the path maximizes $R_{-1/2}$ [12, 125]. The survey [67] further discusses extremal results for $R_{-1/2}(T)$ among different classes of trees:

(a) trees with given number of vertices and leaves;

(b) trees with given diameter and number of vertices;

(c) chemical trees with given number of vertices;

(d) chemical trees with given number of vertices and leaves.

As an example of other related work on this subject, the three largest possible and the three smallest possible values of $R_{-1/2}(T)$, where T is a chemical tree with given number of vertices, are presented in [40]. The extremal trees achieving these extremal values are also characterized.

Many extremal results on $R_{-1/2}(\cdot)$ can be extended to the generalized Randić index $R_\alpha(\cdot)$. It was first shown in [54] that among trees of order $n \geq 5$, $R_\alpha(\cdot)$ is minimized by

- the path for positive α;

- the star for negative α.

The essence of the proof of these results is in fact very similar to that of Proposition 3.1.2 and we encourage the reader to attempt them before checking the reference.

Maximizing the generalized Randić index turns out to be a lot more complicated. For trees, this is studied in several individual cases. In particular, in the aforementioned survey [67] this problem was considered for various classes of trees and chemical trees.

Starting from the next section, we will see many other degree-based indices, which are just examples of numerous such concepts. As an example of how complicated such indices can be, the *higher order Randić index* was proposed in [63] as

$$^iR(G) = \sum_{v_0 v_1 \ldots v_i} \frac{1}{\sqrt{\deg_G(v_0)\deg_G(v_1)\cdots\deg_G(v_i)}},$$

where the sum is taken over paths $v_0, v_1, \ldots, v_i$ of length i in the graph G. It was shown in [122] that among trees on n vertices, the star maximizes the second order Randić index, and among trees on n vertices with maximum degree 3, the path is the unique tree that minimizes the second order Randić index.

3.2   Degree-based indices in trees with a given degree sequence

As in the case of distance-based indices, trees with a given degree sequence and even more general versions of degree-based indices are of interest. We will replace $(\deg_G(u)\deg_G(v))^\alpha$ by a symmetric function $f(\deg_G(u), \deg_G(v))$ (so that the Randić index, along with many other degree-based indices, occurs as a special case) and examine the extremal problems for trees with a given degree sequence.

First, in addition to the Randić index and its generalized version, we list some of the best known indices defined on vertex degrees.

A natural variation of $R(T)$ was named the *sum-connectivity index*, where instead of the product we take the sum of adjacent vertex degrees,

$$\chi(T) = \sum_{uv \in E(T)} (\deg(u) + \deg(v))^{-\frac{1}{2}},$$

and the *general sum-connectivity index*, where the power $-\frac{1}{2}$ is replaced with α as in the case of the Randić index,

$$\chi_\alpha(T) = \sum_{uv \in E(T)} (\deg(u) + \deg(v))^\alpha.$$

Another variant of $R(T)$ is the *harmonic index*

$$H(T) = \sum_{uv \in E(T)} \frac{2}{\deg(u) + \deg(v)},$$

which takes the sum of the reciprocal of the arithmetic mean (as opposed to the geometric mean in the case of $R(T)$) of adjacent vertex degrees.

Another special case of the general sum-connectivity index, with $\alpha = 2$, is the *third Zagreb index*. It is defined as

$$\sum_{uv \in E(T)} (\deg(u) + \deg(v))^2.$$

A slight variant of the third Zagreb index is the *reformulated Zagreb index*, defined as

$$\sum_{uv \in E(T)} (\deg(u) + \deg(v) - 2)^2.$$

Last but certainly not least, the *Atom-Bond connectivity index* [28], defined as

$$\sum_{uv \in E(T)} \sqrt{\frac{\deg(u) + \deg(v) - 2}{\deg(u)\deg(v)}},$$

is a rather complicated example of such graph invariants that has recently received much attention.

These indices, the Randić index, and their generalizations, share the common feature that they are defined on adjacent vertex degrees in graphs/trees. In order to deal with such indices through a unified approach, we first introduce a symmetric bivariate function $f(x, y)$ (defined on $\mathbb{N} \times \mathbb{N}$) such that

$$f(x, a) + f(y, b) \geq f(y, a) + f(x, b) \text{ for any } x \geq y \text{ and } a \geq b. \tag{3.1}$$

Furthermore, strict inequality is implied if both conditions are strict. For a tree T, let the *connectivity function* associated with f be

$$R_f(T) = \sum_{uv \in E(T)} f(\deg(u), \deg(v)). \tag{3.2}$$

Noting that (3.1) is essentially a discrete version of

$$\frac{\partial^2}{\partial x \partial y} f(x, y) \geq 0,$$

it is not difficult to see that with different f, $R_f(T)$ describes $H(T)$, $w_\alpha(T)$ for any α, and $\chi_\alpha(T)$ for $\alpha > 1$ or $\alpha < 0$. We will show that, among trees of given degree sequence, $R_f(T)$ is maximized by the greedy trees (see the previous chapter) and minimized by the so-called *alternating greedy trees*, defined as follows.

Definition 3.2.1 (Alternating greedy trees) *Given the non-increasing degree sequence $(d_1, d_2, \ldots, d_m)$ of internal vertices, an alternating greedy tree is constructed through the following recursive algorithm:*

(i) If $m - 1 \leq d_m$, then the alternating greedy tree is simply obtained by a tree rooted at r with d_m children, $d_m - m + 1$ of which are leaves and the rest with degrees $d_1, \ldots, d_{m-1}$;

(ii) Otherwise, $m - 1 \geq d_m + 1$. We produce a subtree T_1 rooted at r with $d_m - 1$ children with degrees $d_1, \ldots, d_{d_m-1}$;

(iii) Consider the alternating greedy tree S with degree sequence $(d_{d_m}, \ldots, d_{m-1})$, let v be a leaf with the smallest neighbor degree. Identify the root of T_1 with v.

As an example (Figures 3.3, 3.4, and 3.5), for the given degree sequence

$$(8, 7, 6, 6, 5, 5, 3, 3, 3, 2),$$

- T_1 is constructed with degrees $\{8, 2\}$ (as in (ii)), leaving the degree sequence $(7, 6, 6, 5, 5, 3, 3, 3)$ (as in (iii)) with the corresponding alternating greedy tree S_1;

- To construct S_1, T_2 is formed with degrees $\{7, 6, 3\}$, leaving the degree sequence $(6, 5, 5, 3, 3)$ with the corresponding alternating greedy tree S_2;

- To construct S_2, T_3 is formed with degrees $\{6, 5, 3\}$, leaving the degree sequence $(5, 3)$ to provide us the trivial S_3 (as in (i));

- Attaching T_3 to S_3 (i.e., identifying the root of T_3 with a leaf of S_3 whose neighbor has the smallest degree in S_3, as in (iii)) yields S_2;

- Then attaching T_2 to S_2 (i.e., identifying the root of T_2 with a leaf of S_2 whose neighbor has the smallest degree in S_2) yields S_1;

- In the final step, it is obvious that the two choices (two leaves of S_1 with the same neighbor degree) for attaching T_1 to S_1 yield two different alternating greedy trees. Consequently, unlike the greedy trees, alternating greedy trees are not necessarily unique.

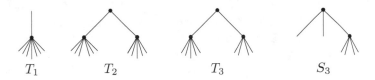

FIGURE 3.3
Construction of T_1, T_2, T_3 and S_3.

FIGURE 3.4
The alternating greedy tree S_1 from T_2, T_3 and S_3.

FIGURE 3.5
The alternating greedy trees T or T' from T_1 and S_1.

Theorem 3.2.1 *For any function f satisfying* (3.1) *and $R_f(T)$ defined as in* (3.2), *$R_f(T)$ is maximized by the greedy tree and minimized by an alternating greedy tree among trees with a given degree sequence.*

3.2.1   Greedy trees

In this section we prove a fundamental property of the extremal trees with a given degree sequence which maximize $R_f(T)$. This property, together with the previously established necessary and sufficient conditions of greedy trees, provide the proof for the extremality of the greedy tree with respect to general $R_f(T)$.

For this purpose, we consider a maximal path (i.e., a path that cannot be extended, not necessarily of maximum length) $P(v_0, v_{t+1})$ in the extremal tree T, with vertices labeled $v_0, v_1, \ldots, v_t, v_{t+1}$, with v_0 and v_{t+1} being leaves. Let T_i ($i = 1, \ldots, t$) denote the connected components containing v_i in $T - E(P(v_0, v_{t+1}))$. As we have already seen in some elementary proofs, we only need to focus on the contribution from each edge to the value of $R_f(T)$. Note that the order of T_i's does not affect this contribution of any edge not on $P(v_0, v_{t+1})$.

Lemma 3.2.1 *Among the extremal trees with a given degree sequence that maximize $R_f(T)$ and have a maximal path $P(v_0, v_{t+1})$ of length $t + 1$ (as defined above), there is one that satisfies (with appropriate labeling), for any given $s \leq (t+1)/2$,*

$$\deg(v_s) \leq \deg(v_{t+1-s}) \leq \deg(v_\ell) \tag{3.3}$$

for every ℓ such that $s \leq \ell \leq t + 1 - s$.

Proof:

Let v_k be the vertex with the largest degree on this path. Without loss of generality, one can assume that

$$\deg(v_{k-1}) \leq \deg(v_{k+1}) \leq \deg(v_k).$$

To prove (3.3), first note that it is sufficient to establish

$$\deg(v_{k-i}) \leq \deg(v_{k+i})$$

and

$$\deg(v_{k+i}) \geq \deg(v_{k+i+1})$$

for all i. Furthermore, this will automatically place v_k as the middle vertex of the path $P(v_0, v_{t+1})$.

Supposing (for contradiction) that (3.3) does not hold, we consider cases where each of the above statements fails:

- Let i be the smallest value such that

$$\deg(v_{k-i}) \leq \deg(v_{k+i})$$

does not hold. Then, we have

$$\deg(v_{k-i}) > \deg(v_{k+i}) \text{ and } \deg(v_{k-i+1}) \leq \deg(v_{k+i-1}).$$

Consider the tree

$$T' = T - v_{k-i}v_{k-i+1} - v_{k+i}v_{k+i-1} + v_{k+i}v_{k-i+1} + v_{k-i}v_{k+i-1}$$

obtained from T by "switching" the two "tails" as in Figure 3.6. From T to T', the values of $f(\cdot, \cdot)$ stay the same for all pairs of adjacent vertex degrees except for the pairs $\{v_{k-i}, v_{k-i+1}\}, \{v_{k+i}, v_{k+i-1}\}$ in T and $\{v_{k+i}, v_{k-i+1}\}, \{v_{k-i}, v_{k+i-1}\}$ in T'. By the condition on $f(\cdot, \cdot)$, we have

$$f(\deg(v_{k-i+1}), \deg(v_{k+i})) + f(\deg(v_{k+i-1}), \deg(v_{k-i}))$$
$$\geq f(\deg(v_{k-i+1}), \deg(v_{k-i})) + f(\deg(v_{k+i-1}), \deg(v_{k+i}))$$

and, consequently,

$$R_f(T') \geq R_f(T).$$

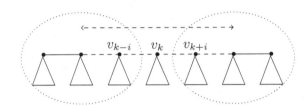

FIGURE 3.6
Illustration of Case (1).

- Similarly, let i be the smallest value such that

$$\deg(v_{k+i}) \geq \deg(v_{k+i+1})$$

does not hold. Note that $i \geq 1$. We establish our arguments in two steps:

  - If $\deg(v_{k+i+2}) < \deg(v_{k+i+1})$, consider the tree

    $$T' = T - v_{k+i}v_{k+i-1} - v_{k+i+2}v_{k+i+1} + v_{k+i}v_{k+i+2} + v_{k+i-1}v_{k+i+1}$$

    obtained from T by "switching" T_{k+i} and T_{k+i+1} as in Figure 3.7. The same argument as in Case (1) shows that

    $$R_f(T') > R_f(T),$$

    a contradiction.
  - More generally, if $\deg(v_{k+i+2}) \geq \deg(v_{k+i+1})$, let j be the largest value such that $\deg(v_{k+i+j}) \geq \deg(v_{k+i+j-1})$ (note that, since v_{t+1} is a leaf, we must have $\deg(v_{t+1}) < \deg(v_t)$). Then, consider the tree

    $$T' = T - v_{k+i}v_{k+i-1} - v_{k+i+j}v_{k+i+j+1} + v_{k+i}v_{k+i+j+1} + v_{k+i-1}v_{k+i+j}$$

    obtained from T by "reversing" the branches from T_{k+i} to T_{k+i+j} as in Figure 3.8 and we have

    $$R_f(T') \geq R_f(T).$$

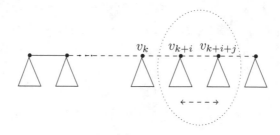

FIGURE 3.7
Illustration of Case (2-a).

FIGURE 3.8
Illustration of Case (2-b).

$\square$

Remark 3.2.1 *We did not need the strictness of inequalities in the above proof as we only intend to show the extremality (but not unique extremality) of the greedy trees.*

Another way to look at this is that we may continue to "switch" or "reverse" different parts of the tree without ever decreasing the value of $R_f(\cdot)$. It is an interesting exercise to show that by repeating the above operations whenever applicable, the number of "contradicting cases" (pairs of degrees that fail our desired property) strictly decreases. Consequently, the entire process terminates after finitely many steps.

For a tree with a given degree sequence where (3.3) holds for every maximal path, it can be shown, through similar arguments as that of Theorem 2.6.3, that it has to be a greedy tree. We leave the details as an exercise. Hence Lemma 3.2.1 implies the extremality of the greedy tree, with respect to $R_f(\cdot)$, among trees with given degree sequence.

3.2.2 Alternating greedy trees

To see that the alternating greedy trees minimize $R_f(T)$ among trees with given degree sequence, we establish characteristics of an extremal tree that minimizes $R_f(T)$, similar to those of Lemma 3.2.1. Again, we start with a maximal path $P(v_0, v_{t+1})$ with vertices $v_0, v_1, \ldots, v_t, v_{t+1}$ in such a tree and claim the following.

Lemma 3.2.2 *Among the extremal trees with a given degree sequence that minimize $R_f(T)$ and have a maximal path $P(v_0, v_{t+1})$ of length $t+1$, there is one that satisfies (with appropriate labeling), for any given $i \leq (t+1)/2$,*

$$\deg(v_i) \leq \deg(v_{t+1-i}) \leq \deg(v_k) \text{ for all } i \leq k \leq t+1-i$$

if i is even; and

$$\deg(v_i) \geq \deg(v_{t+1-i}) \geq \deg(v_k) \text{ for all } i \leq k \leq t+1-i$$

if i is odd.

The proof follows the same lines as that of Lemma 3.2.1. It is also not difficult to see that Lemma 3.2.2 implies the extremality of alternating greedy trees (not necessarily unique). We leave the details of these arguments as exercises.

3.3 Comparison between greedy trees and applications

As in the last section, to facilitate further discussion of these extremal results, we call a symmetric bivariate function $f(x, y)$, defined on $\mathbb{N} \times \mathbb{N}$, *escalating* if (3.1)

$$f(a, b) + f(c, d) \geq f(c, b) + f(a, d) \text{ for all } a \geq c \text{ and } b \geq d,$$

essentially a discrete version of

$$\frac{\partial^2}{\partial x \partial y} f(x, y) \geq 0,$$

is satisfied.

Similarly, a bivariate function $f(x, y)$ defined on $\mathbb{N} \times \mathbb{N}$ is *de-escalating* if

$$f(a, b) + f(c, d) \leq f(c, b) + f(a, d) \text{ for all } a \geq c \text{ and } b \geq d. \quad (3.4)$$

From our proofs in the previous section, the following are immediate consequences.

Theorem 3.3.1 *For any escalating function f and $R_f(T)$ defined as in (3.2), $R_f(T)$ is maximized by the greedy tree among trees with given degree sequence.*

Theorem 3.3.2 *For any de-escalating function f and $R_f(T)$ defined as in (3.2), $R_f(T)$ is minimized by the greedy tree among trees with given degree sequence.*

3.3.1   Between greedy trees

As in the case of the distance-based indices (see in particular Theorem 2.6.5), comparing greedy trees of different degree sequences yields many interesting consequences. With some additional properties one can claim the following for escalating functions. The majority of the related proofs and applications here were introduced in [130].

Theorem 3.3.3 *Given two degree sequences π and π' with $\pi \lhd \pi'$, let T_π^* and $T_{\pi'}^*$ be the greedy trees with degree sequences π and π', respectively. For an escalating function f with*

$$\frac{\partial f}{\partial x} \geq 0 \tag{3.5}$$

and

$$\frac{\partial^2 f}{(\partial x)^2} \geq 0, \tag{3.6}$$

we have

$$R_f(T_\pi^*) \leq R_f(T_{\pi'}^*).$$

Proof:

Given the conditions (3.1), (3.5) and (3.6), we want to show

$$R_f(T_\pi^*) \leq R_f(T_{\pi'}^*)$$

for

$$(d_0, \ldots, d_{n-1}) = \pi \lhd \pi' = (d_0', \ldots, d_{n-1}').$$

By Lemma 1.4.1, we may assume the degree sequences π and π' differ at only two entries, say d_{j_0} (d_{j_0}') and d_{k_0} (d_{k_0}') with $d_{j_0}' = d_{j_0} + 1, d_{k_0}' = d_{k_0} - 1$ for some $j_0 < k_0$.

Let T_π^* contain the vertices u_1 and u_2 with degrees $A := d_{j_0}$ and $C := d_{k_0}$, respectively. Note that from our setup we have $A \geq C$. For convenience, we also introduce the following notations:

- let the parent of u_1 have degree B;

- let the children of u_1 have degrees $B_1, B_2, \ldots, B_{A-1}$;

- let the parent of u_2 have degree D;

- let the children of u_2 have degrees $D_1, D_2, \ldots, D_{C-1}$.

Note that in greedy trees, the vertex degrees are ordered (from largest to smallest) from top to bottom and from one side to another at each level. Thus, we have

$$D \leq B \text{ and } D_i \leq B_j$$

for any $i \in \{1, 2, \ldots, C - 1\}$ and $j \in \{1, 2, \ldots, A - 1\}$.

Pick a child u_3 of u_2, and consider the tree

$$T_{\pi'} = T_\pi^* - u_2 u_3 + u_1 u_3$$

as in Figure 3.9. It is easy to see that $T_{\pi'}$ has degree sequence π', but it is not necessarily a greedy tree.

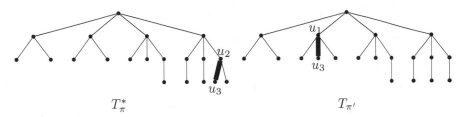

FIGURE 3.9
The trees T_π^* and $T_{\pi'}$ with $\pi = (4, 4, 4, 3, 3, 3, 2, 2, 2, 1, \ldots, 1)$ and $\pi' = (4, 4, 4, 4, 3, 2, 2, 2, 2, 1, \ldots, 1)$.

As before, we only need to focus on the edges whose contribution to $R_f(\cdot)$ changes from T_π^* to $T_{\pi'}$. These are the edges associated with the vertices u_1, u_2 and u_3. Note that the degrees of u_1 and u_2 have changed to $A + 1$ and $C - 1$, respectively.

Considering the edge between u_1 and its parent, we have the change in contribution

$$f(A + 1, B) - f(A, B).$$

Similarly, we have

$$f(C, D) - f(C - 1, D)$$

for the edge between u_2 and its parent. From the edges $u_2 u_3$ and $u_1 u_3$, we have a change in the function value of

$$f(A + 1, D_1) - f(C, D_1).$$

The change in the contributions to the function value from the edges between u_1 and its children can be represented by the sum

$$\sum_{i=1}^{A-1} (f(A + 1, B_i) - f(A, B_i)).$$

Similarly, the change in the contributions to the function value between u_2 and its children can be represented by the sum

$$\sum_{j=2}^{C-1} (f(C, D_j) - f(C - 1, D_j)).$$

Now, we have $R_f(T_{\pi'}) - R_f(T_\pi^*)$ as

$$(f(A+1, D_1) - f(C, D_1)) \tag{3.7}$$
$$+ ((f(A+1, B) - f(A, B)) - (f(C, D) - f(C-1, D))) \tag{3.8}$$
$$+ \left(\sum_{i=1}^{A-1} (f(A+1, B_i) - f(A, B_i)) - \sum_{j=2}^{C-1} (f(C, D_j) - f(C-1, D_j)) \right). \tag{3.9}$$

We now analyze each of the three terms (3.7), (3.8), and (3.9).

- First, we have
$$f(A+1, D_1) - f(C, D_1) \geq 0$$
as $\dfrac{\partial f}{\partial x} \geq 0$ and $A \geq C$.

- Next, note that
$$f(A+1, B) - f(A, B) = \frac{\partial f}{\partial x}(A', B)$$
and
$$f(C, B) - f(C-1, B) = \frac{\partial f}{\partial x}(C', B),$$
where $A \leq A' \leq A+1$ and $C-1 \leq C' \leq C$.

Since $A \geq C$, we have $A' \geq C'$. Then, our assumption $\dfrac{\partial^2 f}{(\partial x)^2} \geq 0$ implies that
$$\frac{\partial f}{\partial x}(A', B) \geq \frac{\partial f}{\partial x}(C', B)$$
and hence
$$f(A+1, B) - f(A, B) \geq f(C, B) - f(C-1, B).$$
Together with
$$(f(C, B) - f(C-1, B)) \geq (f(C, D) - f(C-1, D))$$
(as f is escalating and $C \geq C-1$, $B \geq D$), we have
$$(f(A+1, B) - f(A, B)) - (f(C, D) - f(C-1, D)) \geq 0.$$

- Similarly, we have
$$(f(A+1, B_i) - f(A, B_i)) - (f(C, D_j) - f(C-1, D_j)) \geq 0$$
for any i and j. Hence, every term of
$$\sum_{i=1}^{A-1} (f(A+1, B_i) - f(A, B_i))$$

is larger than every term of

$$\sum_{j=2}^{C-1} (f(C, D_j) - f(C-1, D_j)).$$

Furthermore, note that there are more terms in the first sum than the second since $A - 1 > C - 2$, and that

$$f(A+1, B_i) - f(A, B_i) \geq 0, \qquad f(C, D_j) - f(C-1, D_j) \geq 0$$

for any i, j since $\dfrac{\partial f}{\partial x} \geq 0$.

Consequently, we have

$$\sum_{i=1}^{A-1} (f(A+1, B_i) - f(A, B_i)) - \sum_{j=2}^{C-1} (f(C, D_j) - f(C-1, D_j)) \geq 0.$$

Thus, all three terms (3.7), (3.8) and (3.9) are non-negative. Hence,

$$R_f(T_{\pi'}) - R_f(T_\pi^*) \geq 0.$$

Note that $R_f(T_\pi^*) \geq R_f(T_{\pi'})$ as T_π^* is greedy. Therefore,

$$R_f(T_\pi^*) \leq R_f(T_{\pi'}) \leq R_f(T_{\pi'}^*).$$

$\square$

Remark 3.3.1 *Note that, as in condition (3.1), the discrete version (which is weaker than the continuous version) of the conditions (3.5) and (3.6) would be sufficient for the above argument. Theorem 3.3.3, however, is stated with $\dfrac{\partial f}{\partial x}$ and $\dfrac{\partial^2 f}{(\partial x)^2}$ in order to facilitate the presentation, as well as to simplify the application of the result.*

It is not difficult to see that the analogous statement holds for de-escalating functions with the corresponding additional conditions. The proof is left as an exercise.

Theorem 3.3.4 *Given two degree sequences π and π' with $\pi \lhd \pi'$, let T_π^* and $T_{\pi'}^*$ be the greedy trees with degree sequences π and π', respectively. For a de-escalating function f with*

$$\frac{\partial f}{\partial x} \leq 0 \tag{3.10}$$

and

$$\frac{\partial^2 f}{(\partial x)^2} \leq 0, \tag{3.11}$$

we have

$$R_f(T_\pi^*) \geq R_f(T_{\pi'}^*).$$

3.3.2 Applications to extremal trees

Assume, for convenience, that the function f is escalating, increasing, and convex. The following statements can be proved in exactly the same way as those for distance-based indices. Of course, it is easy to see the analogous statements for de-escalating functions. We leave the details as exercises.

Corollary 3.3.1 *Among all trees of order n, the star maximizes $R_f(.)$.*

Corollary 3.3.2 *Among all trees of order n with given maximum degree Δ, the greedy tree with degree sequence $(\Delta, \Delta, \ldots, \Delta, m, 1, \ldots, 1)$ (where $1 \leq m \leq \Delta - 1$) maximizes $R_f(.)$.*

As mentioned in Section 2.6.3, this extremal tree is sometimes called a "complete Δ-ary tree" or "good Δ-ary tree". We also have similar results for trees with a given number of leaves, given independence number or matching number.

Corollary 3.3.3 *Among all trees of order n with s leaves, the greedy tree with degree sequence $(s, 2, \ldots, 2, 1)$ (s 1s) maximizes $R_f(.)$. This tree is a starlike tree (see Section 1.3).*

Corollary 3.3.4 *Among all trees of order n with independence number α, the greedy tree with degree sequence $(\alpha, 2, \ldots, 2, 1, \ldots, 1)$ maximizes $R_f(.)$.*

Corollary 3.3.5 *Among all trees of order n with matching number β, the greedy tree with degree sequence $(n - \beta, 2, \ldots, 2, 1, \ldots, 1)$ maximizes $R_f(.)$.*

3.3.3 Application to specific indices

Before ending this section, we verify the conditions for Theorems 3.3.1, 3.3.2, and 3.3.3 for various degree-based indices. Many individual extremal questions are answered as corollaries.

Connectivity index

When $f(x, y) = x^\alpha y^\alpha$, recall that

$$R_f(T) = \sum_{uv \in E(T)} (\deg(u) \deg(v))^\alpha$$

is the generalized Randić index. Considering the case $\alpha > 0$, we have

$$\begin{aligned}
f(a, b) + f(c, d) - f(c, b) - f(a, d) &= a^\alpha b^\alpha + c^\alpha d^\alpha - c^\alpha b^\alpha - a^\alpha d^\alpha \\
&= (a^\alpha - c^\alpha)(b^\alpha - d^\alpha) \\
&\geq 0
\end{aligned}$$

for any $a \geq c$ and $b \geq d$. Thus, $f(x, y)$ is escalating and Theorem 3.3.1 holds.

Similarly, $f(x, y)$ is de-escalating for $\alpha < 0$. Consequently, we immediately have the following.

Theorem 3.3.5 *Among trees with given degree sequence, the connectivity index is maximized (minimized) by the greedy tree for $\alpha > 0$ ($\alpha < 0$).*

Remark 3.3.2 *Furthermore, if $\alpha > 1$, it is easy to verify (3.5) and (3.6). Consequently, Theorem 3.3.3 and the corresponding corollaries hold.*

Not much can be done if $\alpha < 1$, which probably partially explains the complexity of the extremal problems with respect to the Randić index and its generalization.

General sum-connectivity index and the third Zagreb index

When $f(x, y) = (x + y)^{\alpha}$, recall that

$$R_f(T) = \chi_\alpha(T) = \sum_{uv \in E(T)} (\deg(u) + \deg(v))^{\alpha}$$

is the general sum-connectivity index. It is simply the sum-connectivity index when $\alpha = 1$.

First, we show that $\chi_\alpha(T)$ is escalating for $\alpha \geq 1$ and de-escalating for $0 < \alpha < 1$.

Consider $\alpha \geq 1$ and let $a \geq c$ and $b \geq d$. To show that $f(x, y)$ is escalating, it suffices to show that

$$(a + b)^{\alpha} - (c + b)^{\alpha} \geq (a + d)^{\alpha} - (c + d)^{\alpha},$$

which is equivalent to

$$\int_{c+b}^{a+b} \alpha t^{\alpha-1} dt \geq \int_{c+d}^{a+d} \alpha t^{\alpha-1} dt$$

through simple calculus. This can be rewritten as

$$\int_{c}^{a} \alpha(t + b)^{\alpha-1} dt \geq \int_{c}^{a} \alpha(t + d)^{\alpha-1} dt,$$

Since $\alpha \geq 1$, we have

$$\alpha(t + b)^{\alpha-1} \geq \alpha(t + d)^{\alpha-1}$$

for $b \geq d$, so the desired inequality holds. Similarly, if $0 < \alpha < 1$, $f(x, y)$ is de-escalating.

Consequently, we have the following as a corollary to Theorem 3.3.1.

Theorem 3.3.6 *Among trees with given degree sequence, the general sum-connectivity index is maximized (minimized) by the greedy tree for $\alpha \geq 1$ $(0 < \alpha < 1)$.*

Remark 3.3.3 *Furthermore, if $\alpha \geq 0$, it is easy to verify (3.5) and (3.6) for $f(x,y) = (x+y)^{\alpha}$. Therefore, Theorem 3.3.3 (when $\alpha \geq 1$ and $f(x,y)$ is escalating) and the corresponding corollaries apply.*

Remark 3.3.4 *Also note that the third Zagreb index as well as the sum-connectivity index itself are both special cases of the general sum-connectivity index. It is easy to check that Theorems 3.3.1 and 3.3.3 and their consequences hold.*

Reformulated Zagreb index

Although the reformulated Zagreb index, defined as

$$\sum_{uv \in E(T)} (\deg(u) + \deg(v) - 2)^2,$$

is not a special case of the general sum-connectivity index, it is rather obvious that it can be analyzed in very similar ways.

Letting $a \geq c$ and $b \geq d$,

$$(a + b - 2)^2 + (c + d - 2)^2 \geq (c + b - 2)^2 + (a + d - 2)^2$$

is equivalent to

$$2(a - c)(b - d) \geq 0,$$

which holds by our conditions.

Thus $f(x,y)$ is escalating and Theorem 3.3.1 holds.

Theorem 3.3.7 *Among trees with given degree sequence, the reformulated Zagreb index is maximized by the greedy tree.*

Remark 3.3.5 *Furthermore, it is easy to verify (3.5) and (3.6) for $f(x,y) = (x + y - 2)^2$. Hence, Theorem 3.3.3 and its consequences apply.*

Atom-Bond connectivity index

When $f(x,y) = \sqrt{\dfrac{x+y-2}{xy}}$, the Atom-Bond connectivity (ABC) index

$$\sum_{uv \in E(T)} \sqrt{\frac{\deg(u) + \deg(v) - 2}{\deg(u)\deg(v)}}$$

is perhaps one of the most complicated graph invariants defined on adjacent vertex degrees. In what follows, to prove that $f(x,y) = \sqrt{\dfrac{x+y-2}{xy}}$ is de-escalating, we first introduce some related facts.

Lemma 3.3.1 *For all positive integers c and d,*

$$f(c+1, d+1) + f(c, d) \leq f(c, d+1) + f(c+1, d). \qquad (3.12)$$

Proof:

First note that

$$
\left(\frac{1}{c+1} + \frac{1}{d} - \frac{2}{(c+1)d} \right) \left(\frac{1}{c} + \frac{1}{d+1} - \frac{2}{c(d+1)} \right)
$$
$$
- \left(\frac{1}{c} + \frac{1}{d} - \frac{2}{cd} \right) \left(\frac{1}{c+1} + \frac{1}{d+1} - \frac{2}{(c+1)(d+1)} \right)
$$
$$
= \left(\frac{1}{c} - \frac{1}{c+1} \right) \left(\frac{1}{d} - \frac{1}{d+1} \right)
$$
$$
> 0,
$$

so

$$
(f(c, d+1) + f(c+1, d))^2 - (f(c+1, d+1) + f(c, d))^2
$$
$$
= 2\sqrt{\left(\frac{1}{c+1} + \frac{1}{d} - \frac{2}{(c+1)d} \right) \left(\frac{1}{c} + \frac{1}{d+1} - \frac{2}{c(d+1)} \right)}
$$
$$
- 2\sqrt{\left(\frac{1}{c} + \frac{1}{d} - \frac{2}{cd} \right) \left(\frac{1}{c+1} + \frac{1}{d+1} - \frac{2}{(c+1)(d+1)} \right)}
$$
$$
+ \frac{2}{cd(c+1)(d+1)}
$$
$$
> 0.
$$

Thus, (3.12) follows.

$\square$

Lemma 3.3.2 *For any non-negative integer k and positive integers c, d,*

$$f(c+k, d+1) + f(c, d) \leq f(c, d+1) + f(c+k, d). \qquad (3.13)$$

Proof:

Through repeated applications of (3.12), we have

$$
\begin{aligned}
f(c+k, d+1) - f(c+k, d) &\leq f(c+k-1, d+1) - f(c+k-1, d) \\
&\leq f(c+k-2, d+1) - f(c+k-2, d) \\
&\leq \cdots \\
&\leq f(c, d+1) - f(c, d).
\end{aligned}
$$

$\square$

We can now show that the corresponding function is indeed de-escalating.

Proposition 3.3.1 $f(x,y) = \sqrt{\dfrac{x+y-2}{xy}}$ *de-escalating on* $\mathbb{N} \times \mathbb{N}$.

Proof:

By definition, we want to show

$$f(a,b) + f(c,d) \le f(c,b) + f(a,d)$$

for any $a \ge c$ and $b \ge d$.

Let $a = c + k$ and $b = d + r$ with non-negative integers k, r. Through repeated applications of (3.13), we have

$$
\begin{aligned}
f(a,b) - f(c,b) &= f(c+k, d+r) - f(c, d+r) \\
&\le f(c+k, d+r-1) - f(c, d+r-1) \\
&\le f(c+k, d+r-2) + f(c, d+r-2) \\
&\le \cdots \\
&\le f(c+k, d) - f(c,d) \\
&= f(a,d) - f(c,d).
\end{aligned}
$$

$\square$

Hence, by Proposition 3.3.1 and Theorem 3.3.1, we have the following theorem.

Theorem 3.3.8 *Among trees with given degree sequence, the Atom-Bond connectivity (ABC) index is minimized by the greedy tree.*

Remark 3.3.6 *Although the greedy tree is indeed extremal, unfortunately, (3.10) and (3.11) do not both hold in order to apply Theorem 3.3.4. This may also explain the lack of "simple" extremal results with respect to the ABC index, which we will further explore later in this chapter.*

3.4   The Zagreb indices

Besides the Randić index and other degree-based indices defined on adjacent vertex degrees, the sum of the squares of degrees was probably one of the first graph indices defined on degrees. Given a tree T (or a general graph),

$$M_1(T) = \sum_{v \in V(T)} (\deg_T(v))^2$$

is called the first Zagreb index [44]. Also, recall that a special case of R_α (when $\alpha = 1$) is the second Zagreb index [44]

$$M_2(T) = \sum_{uv \in E(T)} (\deg_T(u) \cdot \deg_T(v)).$$

We have already examined the extremal problems related to $M_2(\cdot)$ as a special case of R_α. Among general graphs on n vertices, it is easy to see that adding an edge will increase the degrees of corresponding vertices and hence the value of $M_1(\cdot)$. Hence, we have the following.

Proposition 3.4.1 *Among connected graphs of order n, the first Zagreb index is maximized by the complete graph K_n and minimized by some tree.*

Among trees of a given degree sequence, it is easy to see that $M_1(\cdot)$ is a constant. Between different degree sequences simple algebra shows the following (much simpler) analogue of Theorem 3.3.3. We leave the proof as an exercise.

Proposition 3.4.2 *Given two degree sequences π and π' with $\pi \lhd \pi'$. Let T and T' be two trees with degree sequences π and π', respectively. Then,*

$$M_1(T') > M_1(T).$$

As a consequence, by finding the degree sequence that majorizes, or is majorized by, all other degree sequences of trees on n vertices, and the unique trees with the extremal degree sequences, we have the following.

Proposition 3.4.3 *If T is a tree on n vertices, then*

$$4n - 6 \leq M_1(T) \leq n(n-1)$$

with left equality if and only if T is the path P_n and right equality if and only if T is the star $K_{1,n-1}$.

In addition to the numerous common problems in chemical graph theory (as we have already seen in the previous sections), it is also natural to examine how different or similar M_1 and M_2 can be among graphs and especially trees. This was examined in detail in [116], on which the majority of this section is based.

3.4.1 Graphs with $M_1 = M_2$

There are many graphs for which $M_1(G) = M_2(G)$. For example, one can easily verify that

$$M_1(C_n) = M_2(C_n)$$

in a cycle C_n.

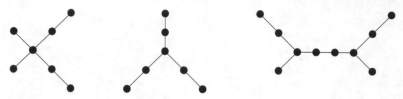

FIGURE 3.10
Trees with $M_1(T) = M_2(T)$.

Among trees, it may be less obvious to find such examples. Figure 3.10 shows a few small trees where $M_1(T)$ and $M_2(T)$ share the same value.

A natural question that follows is whether there are infinitely many trees where the Zagreb indices share the same value. The following observation provides a positive answer to this question.

Proposition 3.4.4 *Let T be a tree with two adjacent vertices u, v, each of degree 2, and let T' be obtained from T through subdividing the edge uv (i.e., adding a new vertex w that is adjacent to both u and v, and removing the edge uv, see Figure 3.11). Then*

$$M_2(T') - M_1(T') = M_2(T) - M_1(T). \qquad (3.14)$$

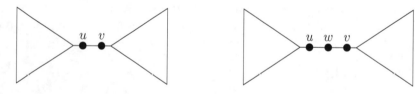

FIGURE 3.11
Subdividing T (on the left) to obtain T' (on the right).

Proof:

Note that the degrees of u and v stay as 2 from T to T', so we have

$$M_1(T') - M_1(T) = (\deg(w))^2 = 4$$

and

$$M_2(T') - M_2(T)$$
$$= \deg_{T'}(w) \deg_{T'}(u) + \deg_{T'}(w) \deg_{T'}(v) - \deg_T(v) \deg_T(u)$$
$$= 4.$$

□

Remark 3.4.1 *The above statement presents an operation that maintains the difference between the Zagreb indices while generating a new tree of larger order. Therefore, starting from one of the examples in Figure 3.10 and repeating this operation will result in infinitely many non-isomorphic trees on which the Zagreb indices share the same value.*

Although this is only stated and illustrated for trees, it is not difficult to see that the same holds for general graphs. The proof is exactly the same, and we leave it as an exercise.

Along the same line, there are also many other conditions that result in (3.14) for two different trees T and T' of the same order. The following is a special case that will be useful in other related studies.

Proposition 3.4.5 *Let T be a tree with vertices u, v, w such that v is the only non-leaf neighbor of u and w, respectively, and let T' be obtained from T through detaching any number of pendant edges from w and reattaching them to u (Figure 3.12). Then*

$$M_2(T') - M_1(T') = M_2(T) - M_1(T).$$

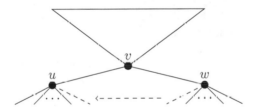

FIGURE 3.12
"Moving" pendant edges from w to u in T to obtain T'.

Proof:

It suffices to consider the case of "moving" one pendant edge from w to u as in Figure 3.12. For convenience let

$$\deg_T(v) = d_1, \ \deg_T(u) = d_2 \text{ and } \deg_T(w) = d_3,$$

then

$$\deg_{T'}(u) = d_2 + 1 \text{ and } \deg_{T'}(w) = d_3 - 1.$$

Noting that, from T to T', u and w are the only vertices whose degree changed, we immediately have

$$M_1(T') - M_1(T) = (d_2 + 1)^2 + (d_3 - 1)^2 - d_2^2 - d_3^2 = 2d_2 - 2d_3 + 2.$$

Dealing with M_2, only the edges incident with u and w need to be considered. Direct calculation yields

$$
\begin{aligned}
M_2(T') &- M_2(T) \\
&= d_1(d_2+1) + d_1(d_3-1) + d_2(d_2+1) + (d_3-2)(d_3-1) \\
&\qquad - d_1 d_2 - d_1 d_3 - d_2(d_2-1) - d_3(d_3-1) \\
&= 2d_2 - 2d_3 + 2.
\end{aligned}
$$

□

3.4.2   Maximum $M_2(\cdot) - M_1(\cdot)$ in trees

Next, we will examine the difference between $M_2(\cdot)$ and $M_1(\cdot)$, starting with $M_2(\cdot) - M_1(\cdot)$.

Let us first consider trees with a given degree sequence $(d_1, d_2, \ldots)$ and hence some given order n. Since $M_1(\cdot)$ is simply a constant (the sum of squares of fixed values) and M_2 is maximized by the greedy tree with the given degree sequence, we only need to consider greedy trees (of different degree sequences) in order to maximize $M_2(\cdot) - M_1(\cdot)$.

We first show that to find the maximum $M_2(\cdot) - M_1(\cdot)$ among trees of given order n, we only need to consider trees of small diameter and "centered" at the vertex with the largest degree.

Lemma 3.4.1 *For every fixed number of vertices n, there exists an extremal tree that maximizes $M_2(\cdot) - M_1(\cdot)$ with diameter at most 4 and all leaves at most distance 2 away from the vertex v of the largest degree.*

Proof:

Without loss of generality, we may assume such an extremal tree to be a greedy tree T rooted at $v = v_1$ with degree sequence

$$
(d_1, d_2, \ldots, d_{d_1+1}, \ldots)
$$

where $d_2 \geq d_3 \geq \ldots \geq d_{d_1+1}$ are the degrees of the children of v, labeled as $v_2, v_3, \ldots, v_{d_1+1}$.

Suppose that at least one of $v_2, v_3, \ldots, v_{d_1+1}$ has a non-leaf child, let this vertex be v_2 (without loss of generality, since the tree is greedy) with a non-leaf child $v_{2,1}$. Further assume that v_2 has children $v_{2,1}, v_{2,2}, \ldots, v_{2,d_2-1}$ with degrees $e_1 \geq e_2 \geq \ldots \geq e_{d_2-1}$, respectively. See Figure 3.13.

Let us consider two cases depending on the degree of $v_{2,1}$:

- If $e_1 \geq 3$, define

$$
T' = T - v_2 v_{2,1} + v_1 v_{2,1}
$$

as in Figure 3.14. That is, T' is obtained from T by "moving" the subtree induced by $v_{2,1}$ and its descendants from v_2 to v_1.

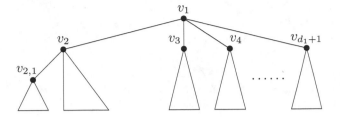

FIGURE 3.13
The greedy tree rooted at v.

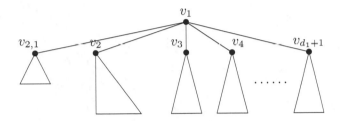

FIGURE 3.14
The tree $T' = T - v_2 v_{2,1} + v_1 v_{2,1}$.

Since v_1 and v_2 are the only vertices with degree changes from T to T', we have

$$M_1(T') - M_1(T) = (d_1 + 1)^2 + (d_2 - 1)^2 - d_1^2 - d_2^2 = 2d_1 - 2d_2 + 2.$$

Considering all the edges incident with v_1 or v_2, we have

$$M_2(T') - M_2(T)$$

$$= (d_2 - 1)(d_1 + 1) + (d_2 - 1)\sum_{i=2}^{d_2-1} e_i + (d_1 + 1)\left(e_1 + \sum_{i=3}^{d_1+1} d_i\right)$$

$$- d_1 d_2 - d_2 \sum_{i=1}^{d_2-1} e_i - d_1 \sum_{i=3}^{d_1+1} d_i$$

$$= -(d_1 - d_2 + 1) + (d_1 - d_2 + 1)e_1 + \sum_{i=3}^{d_1+1} d_i - \sum_{i=2}^{d_2-1} e_i$$

$$> (d_1 - d_2 + 1)(e_1 - 1)$$

$$\geq 2(d_1 - d_2 + 1)$$

$$= M_1(T') - M_1(T)$$

where the first inequality follows from the fact that in this greedy tree we have

$$d_1 \geq d_2 \geq d_3 \geq \ldots \geq d_{d_1+1} \geq e_1 \geq e_2 \geq \ldots \geq e_{d_2-1}.$$

- Now, suppose $e_1 \leq 2$. Since we already assumed that $v_{2,1}$ is not a leaf, we must have $e_1 = 2$. Let the only child of $v_{2,1}$ be u with degree $1 \leq e \leq e_{d_2-1} \leq 2 = e_1$. Consider the tree

$$T'' = T - v_{2,1}u + v_2u$$

as in Figure 3.15, where we simply "detached" the subtree induced by u and its descendants from $v_{2,1}$ and reattached it at v_2.

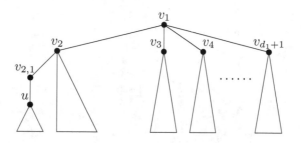

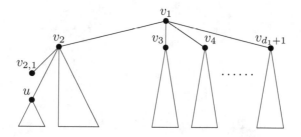

FIGURE 3.15
The greedy tree rooted at v with $\deg(v_{2,1}) = 2$ (top) and the tree T'' (bottom).

Similar arguments as before yield

$$M_1(T'') - M_1(T) = (d_2 + 1)^2 + 1^2 - d_2^2 - 2^2 = 2d_2 - 2$$

and

$$M_2(T'') - M_2(T) = (d_2 + 1)d_1 + (d_2 + 1) + (d_2 + 1)e + (d_2 + 1)\sum_{i=2}^{d_2-1} e_i$$

$$- d_2 d_1 - 2d_2 - 2e - d_2 \sum_{i=2}^{d_2-1} e_i$$

$$= d_1 - d_2 + 1 + (d_2 - 1)e + \sum_{i=2}^{d_2-1} e_i$$

$$\geq d_1 + d_2 - 2 \geq 2d_2 - 2$$

$$= M_1(T'') - M_1(T)$$

where the first inequality follows from the facts that $e \geq 1$ and $e_i \geq 1$ for $2 \leq i \leq d_2 - 1$.

Note that in either case, the new tree T' or T'' has at least as large a value of $M_2(\cdot) - M_1(\cdot)$ as T and yields a new degree sequence that strictly "majorizes" the original (recall Definition 1.4.1). Hence we can continue this operation (whenever $v_{2,1}$ in the current tree is not a leaf), rearranging the tree to be greedy (which can only increase the value of $M_2(\cdot) - M_1(\cdot)$) at each step, and the process will terminate. This results in an extremal tree with maximum $M_2(\cdot) - M_1(\cdot)$ where all leaves are at most distance 2 away from the vertex of the largest degree. □

Remark 3.4.2 *In fact, a stronger statement holds than that of Lemma 3.4.1, namely that any extremal tree with the maximum $M_2(\cdot) - M_1(\cdot)$ cannot have diameter more than 4. This can be seen by considering the second case of the above proof, when $e_1 = 2$,*

$$M_2(T'') - M_2(T) = M_1(T'') - M_1(T)$$

only if $e = e_i = 1$ for any $2 \leq i \leq d_2 - 1$ and $d_1 = d_2$. In this case, $\deg_{T''}(v_2) > \deg_{T''}(v_1)$, so rearranging T'' to be a greedy tree and continuing the described operations will strictly increase $M_2(\cdot) - M_1(\cdot)$. We skip these details (which we leave as an exercise) as Lemma 3.4.1 is sufficient for our purpose.

Analogous to Lemma 3.4.1 and Proposition 3.4.5, we will show that one may decrease the diameter to 3 without decreasing the value of $M_2(\cdot) - M_1(\cdot)$. Thus, we have the following.

Theorem 3.4.1 *There exists an extremal tree that maximizes $M_2(\cdot) - M_1(\cdot)$ with diameter at most 3.*

Proof:

Suppose, without loss of generality, that a tree is the result of Lemma 3.4.1 with root v and children $v_2, v_3, \ldots$ (each of which has only leaf children). By Proposition 3.4.5, we may "move" any pendant edge from other internal vertices (except v) to v_2 (without changing the value of $M_2(\cdot) - M_1(\cdot)$). Repeatedly doing so yields a tree with only two internal vertices v and v_2. We leave the details as exercises. □

Next, let us examine this maximum value. According to the above discussion, let v_1, v_2 be the only internal vertices in the tree T with maximum $M_2(\cdot) - M_1(\cdot)$ with degrees d_1 and d_2, respectively. Assume, without loss of generality, that $d_1 \geq d_2$.

If $d_1 > d_2 + 1$, consider a new tree T' obtained from T by detaching one pendant edge from v_1 and attaching it to v_2. From T to T', through similar computation as before, we have

$$M_1(T') - M_1(T) = (d_2 + 1)^2 + (d_1 - 1)^2 - d_2^2 - d_1^2 = 2d_2 - 2d_1 + 2$$

and

$$\begin{aligned} M_2(T') - M_2(T) &= (d_2 + 1)(d_1 - 1) + (d_1 - 1)(d_1 - 2) + (d_2 + 1)d_2 \\ &\quad - d_1 d_2 - d_1(d_1 - 1) - d_2(d_2 - 1) \\ &= d_2 - d_1 + 1 \\ &> 2d_2 - 2d_1 + 2 \\ &= M_1(T') - M_1(T), \end{aligned}$$

where the inequality follows from $d_1 > d_2 + 1$. Hence, we have

$$M_2(T') - M_1(T') > M_2(T) - M_1(T),$$

a contradiction to the extremality of T.

Now, we may assume that the maximum $M_2(\cdot) - M_1(\cdot)$ is only achieved when $|d_1 - d_2| \le 1$. For such a tree T of order n, we have:

- for even n,

$$M_1(T) = 2\left(\frac{n}{2}\right)^2 + n - 2 = \frac{n^2}{2} + n - 2$$

  and

$$M_2(T) = \left(\frac{n}{2}\right)^2 + \frac{n}{2} \cdot 2\left(\frac{n}{2} - 1\right) = \frac{3n^2}{4} - n.$$

- for odd n,

$$M_1(T) = \left(\frac{n-1}{2}\right)^2 + \left(\frac{n+1}{2}\right)^2 + n - 2 = \frac{n^2 + 1}{2} + n - 2$$

  and

$$M_2(T) = \frac{n-1}{2} \cdot \frac{n+1}{2} + \frac{n+1}{2} \cdot \frac{n-1}{2} + \frac{n-1}{2} \cdot \frac{n-3}{2} = \frac{3n^2 + 1}{4} - n.$$

Consequently, we have the following.

Theorem 3.4.2 *For any tree T of order n and diameter at least 3,*

$$M_2(T) - M_1(T) \le \left\lfloor \frac{n^2}{4} \right\rfloor - 2n + 2.$$

Remark 3.4.3 *It is easy to see from the proofs of the above statements that lead to Theorem 3.4.2 that the upper bound is sharp and there is more than one tree that achieves this upper bound.*

We now consider the trees with diameter 2, i.e., when T is a star. It is easy to compute the value of $M_2(\cdot) - M_1(\cdot)$ as the following. We leave the details as exercises.

Proposition 3.4.6 *For a star $K_{1,n-1}$ on $n \geq 3$ vertices, we have*

$$M_2(K_{1,n-1}) - M_1(K_{1,n-1}) = (n-1)(n-1) - \left((n-1)^2 + n - 1\right)$$
$$= -(n-1).$$

From Theorems 3.4.1 and 3.4.2 and Proposition 3.4.6 we may extend the statement of Theorem 3.4.2 to all trees of a given order.

Corollary 3.4.1 *For all trees T of order n,*

$$M_2(T) - M_1(T) \leq \left\lfloor \frac{n^2}{4} \right\rfloor - 2n + 2.$$

Proof:

Note that when $n \geq 4$,

$$\left\lfloor \frac{n^2}{4} \right\rfloor - n + 1 > \frac{n^2}{4} - 1 - n + 1 = \left(\frac{n}{2} - 1\right)^2 - 1 \geq 0,$$

implying that

$$\left\lfloor \frac{n^2}{4} \right\rfloor - 2n + 2 \geq -(n-1). \tag{3.15}$$

It is easy to verify (3.15) for $n = 2, 3$. □

3.4.3 Maximum $M_1(\cdot) - M_2(\cdot)$ in trees

Now, to explore the other extremal end, Proposition 3.4.6 implies that

$$M_1(K_{1,n-1}) - M_2(K_{1,n-1}) = n - 1 \geq 2$$

for stars of order $n \geq 3$.

On the other hand, for a tree that is not a star (or, equivalently, a tree with diameter at least 3), the following theorem states the interesting fact that $M_1(\cdot)$ can never be too much larger than $M_2(\cdot)$. Perhaps even more counter-intuitively, the second largest possible value of $M_1 - M_2$ is achieved by a path as shown below.

Theorem 3.4.3 *Among all trees with diameter at least 3, the path achieves the maximum value 2 of $M_1(\cdot) - M_2(\cdot)$.*

Remark 3.4.4 *It is rather unexpected that the maximum value of $M_2(\cdot) - M_1(\cdot)$ and $M_1(\cdot) - M_2(\cdot)$ are both achieved by trees with small diameters while the second largest $M_1(\cdot) - M_2(\cdot)$ is achieved by the path, commonly considered as the "opposite" structure of a star.*

Proof:

It is easy to compute, for a path P_n of order n, that

$$M_1(P_n) = 4n - 6 \text{ and } M_2(P_n) = 4n - 8,$$

so $M_1(P_n) - M_2(P_n) = 2$.

Now consider an extremal tree T that maximizes $M_1(\cdot) - M_2(\cdot)$ among trees with n vertices whose diameter is greater than 2. Suppose (for contradiction) that

$$M_1(T) - M_2(T) > 2,$$

thus in particular that T is not a path. Let $v_0, v_1, v_2, \ldots, v_{k-1}, v_k, v_{k+1}$ with $k \geq 2$ be a longest path of T. Some internal vertex of the longest path has neighbors not on this path. Without loss of generality, we may also assume that v_0 is closer than v_{k+1} to such a vertex and let v_i be the closest such vertex to v_0. We can also assume that i is minimal among all extremal trees.

In other words, all vertices

$$v_1, v_2, \ldots, v_{i-1} \text{ and } v_{k-i+2}, \ldots, v_k$$

have degree 2 and v_i has degree at least 3 (i.e., v_i has a neighbor not on the path $v_0, v_1, v_2, \ldots, v_{k-1}, v_k, v_{k+1}$). For convenience, let $d_i = \deg(v_i)$ for $1 \leq i \leq k$. We now consider two cases.

- If $i \geq 2$ (i.e., the neighbor of v_0 is of degree 2), then by our assumption we also have $i \leq k - 1$. Let T' be the tree obtained from T by "shifting" the pendant branch at v_i to v_{i-1} (Figure 3.16). This can also be formally understood as "reversing" the edge $v_{i-1}v_i$ and the corresponding branches in T, yielding

$$T' = T - v_{i-2}v_{i-1} - v_i v_{i+1} + v_{i-2}v_i + v_{i+1}v_{i-1}$$

with the switch of labeling of v_{i-1} and v_i.

FIGURE 3.16
The trees T (on the left) and T' (on the right).

It is easy to see that the degree sequence of T' is exactly the same as that of T, hence $M_1(\cdot)$ stays the same from T to T'. For $M_2(\cdot)$, the only edges we need to consider (i.e., those incident to vertices whose degrees changed) are $v_{i-2}v_{i-1}$, $v_{i-1}v_i$ and v_iv_{i+1}. Therefore, we have

$$
\begin{aligned}
M_2(T') - M_2(T) &= d_i(d_{i-2} + d_{i-1}) + d_{i-1}d_{i+1} - d_i(d_{i-1} + d_{i+1}) - d_{i-2}d_{i-1} \\
&= d_id_{i-2} + d_{i-1}d_{i+1} - d_id_{i+1} - d_{i-2}d_{i-1} \\
&= (d_i - d_{i-1})(d_{i-2} - d_{i+1}) \\
&\leq 0 \\
&= M_1(T') - M_1(T),
\end{aligned}
$$

where the inequality follows from the facts that $d_{i-2} \leq d_{i-1} = 2$, $d_{i+1} \geq 2$ and $d_i \geq 3$. Consequently,

$$
M_1(T') - M_2(T') \geq M_1(T) - M_2(T),
$$

with equality if and only if $d_{i-2} = d_{i+1} = 2$. If strict inequality holds, we obtain an immediate contradiction. Otherwise, T' is another extremal tree, which contradicts the minimality of i in our choice of T. In either case, we are done.

- We now assume that $i = 1$. Since $v_0, \ldots, v_{k+1}$ is a longest path, all but one neighbor of v_1 are leaves. Let T'' be the tree obtained from T by detaching all pendant edges at v_1 except v_0v_1 and reattaching them at v_0. See Figure 3.17.

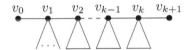
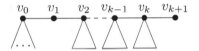

FIGURE 3.17
The trees T (on the left) and T'' (on the right).

Similar to the arguments before, the only vertices with degree changes from T to T'' are v_0 and v_1. Consequently, we have

$$
M_1(T'') - M_1(T) = (d_1 - 1)^2 + 2^2 - d_1^2 - 1 = -2(d_1 - 2)
$$

and

$$
\begin{aligned}
M_2(T'') - M_2(T) &= 2(d_1 - 1) + (d_1 - 1)(d_1 - 2) + 2d_2 \\
&\quad - d_1 - d_1(d_1 - 2) - d_1d_2 \\
&= -(d_1 - 2)d_2 \\
&\leq -2(d_1 - 2) \\
&= M_1(T'') - M_1(T)
\end{aligned}
$$

where the inequality follows from $d_2 \geq 2$ and $d_1 \geq 3$.

Here, for a tree T that is not a path, the argument in the first case shows that we must have $d_1 \geq 3$. Then, applying the second case yields a new tree with larger diameter. This process terminates when a path is achieved, while at every step the value of $M_1(\cdot) - M_2(\cdot)$ can only increase. Hence, the maximum of $M_1(\cdot) - M_2(\cdot)$ is 2 and it is independent of the order of the tree.   □

Remark 3.4.5 *It is obvious that the extremal trees in the above case are not unique, and it is also not difficult to construct (based on the arguments in the proof) other trees that achieve $M_1(\cdot) - M_2(\cdot) = 2$. Such trees include (but are not limited to) well-known structures such as the comet or dumbbell. See Figure 3.18. It is a good exercise to verify $M_1(\cdot) - M_2(\cdot) = 2$ for these trees.*

FIGURE 3.18
Some trees T with $M_1(T) - M_2(T) = 2$.

3.4.4   Further analysis of the behavior of $M_1(\cdot) - M_2(\cdot)$

With a good understanding of the difference between the Zagreb indices, it is a natural question to ask which values this difference can take on and which it cannot. First of all, there are indeed trees T such that $M_1(T) - M_2(T) = 1$. For instance, a tree obtained from attaching a pendant edge to any internal vertex (but not the ones adjacent to the ends) of a path (Figure 3.19).

FIGURE 3.19
A tree T with $M_1(T) - M_2(T) = 1$.

From Theorem 3.4.3 and information from previous sections, we may state the following in regard to an "inverse" question, as in the case of the Wiener index.

Corollary 3.4.2 *There exists a tree with $M_1 - M_2 = x$ for any non-negative integer x.*

Another observation is that there seem to be fewer trees with positive $M_1(\cdot) - M_2(\cdot)$ than those with negative $M_1(\cdot) - M_2(\cdot)$. To formally verify this, we first consider a sufficient condition for $M_1(\cdot) - M_2(\cdot) < 0$ for a tree T.

Theorem 3.4.4 *If there is an edge uv with $\deg(u) \geq 3$ and $\deg(v) \geq 3$ in a tree T, and if neither u nor v is adjacent to any leaf, then we must have*

$$M_1(T) - M_2(T) < 0.$$

Proof:

Let T be such a tree and suppose $M_1(T) - M_2(T)$ is maximum among all such trees. For convenience let $d_1 = \deg(u)$ and $d_2 = \deg(v)$, and let $u_1, u_2, \ldots, u_{d_1-1}$ be the neighbors of u other than v and $v_1, v_2, \ldots, v_{d_2-1}$ be the neighbors of v other than u. Lastly, let T_{u_i} and T_{v_j} denote the connected components containing u_i and v_j, respectively, in $T - \{u, v\}$, for $1 \leq i \leq d_1 - 1$ and $1 \leq j \leq d_2 - 1$ (Figure 3.20).

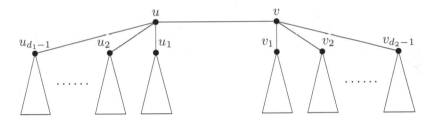

FIGURE 3.20
The tree T with u, v and their neighbors.

We claim that each T_{u_i} or T_{v_j} can be replaced by a path without decreasing the value of $M_1(T) - M_2(T)$. This can be seen by taking the longest path from u (v) to a leaf in T_{u_i} (T_{v_j}) and applying exactly the same arguments of Theorem 3.4.3. Such a path can be extended (if T_{u_i} or T_{v_j} is not a path) without decreasing $M_1(\cdot) - M_2(\cdot)$ until T_{u_i} (T_{v_j}) is a path. We leave the detailed arguments as exercises.

Further, note that each of these paths T_{u_i} or T_{v_j} is of length at least 2 as u and v are not adjacent to leaves.

The structure of T is specifically presented in Figure 3.21. It is easy to compute the values of the Zagreb indices directly as

$$M_1(T) = d_1^2 + d_2^2 + (d_1 + d_2 - 2) + 4(n - d_1 - d_2)$$

and

$$M_2(T) = d_1 d_2 + 2d_1(d_1 - 1) + 2d_2(d_2 - 1) + 2(d_1 + d_2 - 2) + 4(n - 2d_1 - 2d_2 + 2).$$

Consequently, we have

$$
\begin{aligned}
M_2(T) - M_1(T) &= d_1^2 + d_2^2 + d_1 d_2 - 5d_1 - 5d_2 + 6 \\
&= d_1(d_1 - 3) + d_2(d_2 - 3) + (d_1 - 2)(d_2 - 2) + 2 \\
&> 0
\end{aligned}
$$

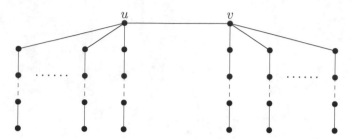

FIGURE 3.21
An extremal tree with maximum $M_1 - M_2$.

as $d_1, d_2 \geq 3$.   □

Recall the classic fact that the probability of a random tree of order n containing any given rooted subtree approaches 1 as $n \to \infty$ [95]. Theorem 3.4.4 implies the following interesting observation.

Corollary 3.4.3 *For large n almost all trees of order n have negative value for $M_1(\cdot) - M_2(\cdot)$.*

3.5   More on the ABC index

Among the degree-based graph indices that have been listed so far, the ABC index is the most complicated. For this reason and for its applications in various molecular graphs, the investigation of the ABC index has been of particular interest. In addition to those on trees that have already been presented, in this section we consider the extremal problems with respect to the ABC index in general graphs. The majority of the content of this section was established in [129].

Throughout this section, a graph that minimizes the ABC index (among a certain category of graphs) is called "optimal" for convenience.

Similar to the case of trees with a given degree sequence, letting $\Gamma(\pi)$ denote the family of connected graphs with degree sequence π, we first derive structural properties of optimal graphs in $\Gamma(\pi)$. These characteristics of the optimal graph are then used to construct a unique special graph, shown to be optimal under certain conditions (Theorem 3.5.1). We will then apply this result to unicyclic and bicyclic graphs, where the corresponding optimal graphs are characterized.

3.5.1 Defining the optimal graph

In order to introduce the structure of the optimal graph and the corresponding theorem, we need to first go over some technical terminologies.

For a graph $G = (V, E)$ with a root v_0, the distance $d(v, v_0)$ is called the *height $h(v)$* of a vertex v, and $h(G) = \max\{h(v)|v \in V\}$ is the height of G. A breadth-first search (BFS) ordering of such a rooted graph was probably first defined in [126] as the following.

Definition 3.5.1 *Let $G = (V, E)$ be a connected rooted graph with a root v_0. A well-ordering $\prec$ of the vertices is called* breadth-first search ordering *with non-increasing degrees (BFS-ordering for short) if the following conditions hold for all vertices $u, v \in V$:*

- $u \prec v$ *implies* $h(u) \leq h(v)$;

- $u \prec v$ *implies* $d(u) \geq d(v)$;

- *Let $uv, xy \in E$ and $uy, xv \notin E$ with $h(u) = h(x) = h(v) - 1 = h(y) - 1$. If $u \prec x$, then $v \prec y$.*

A graph with a BFS-ordering of its vertices is called a *BFS-graph*. If a BFS-graph is a tree, then it is isomorphic to the greedy tree, and also called a *BFS-tree* (see [128]).

It is a good exercise for the interested reader to verify the following useful properties related to the BFS-ordering or BFS-graphs:

- Every graph has an ordering of its vertices which satisfies the conditions (1) and (3) by using breadth-first search. But not all connected graphs have an ordering that satisfy the condition (2). Hence, not all connected graphs are BFS-graphs.

- For a given graphical degree sequence π, there always exists a BFS-graph G in $\Gamma(\pi)$.

We are now ready to construct the optimal graph $G^*(\pi)$ in $\Gamma(\pi)$ for a graphical degree sequence $\pi = (d_1, \ldots, d_n)$ with

$$\sum_{i=1}^{n} d_i = 2(n + c - 1), \; c \geq 0,$$

and $d_1 \geq d_2 \geq c + 1$.

- Select the vertex v_1 with degree d_1 as the root vertex;

- Select the vertices $v_2, v_3, v_4, \ldots, v_{d_1+1}$ with degrees $d_2, \ldots, d_{d_1+1}$ such that

$$N(v_1) = \{v_2, v_3, v_4, \ldots, v_{d_1+1}\};$$

- Append $d_2 - c - 1$ new vertices to v_2, $d_3 - 2$ new vertices to $v_3, \ldots, d_{c+2} - 2$ new vertices to v_{c+2} such that

  - $N(v_2) = \{v_1, v_3, \ldots, v_{c+2}, v_{d_1+2}, v_{d_1+3}, \ldots, v_{d_1+d_2-c}\}$,
  - $N(v_3) = \{v_1, v_2, v_{d_1+d_2-c+1}, \ldots, v_{d_1+d_2+d_3-c-2}\}$,
  - $\cdots$,
  - $N(v_{c+2}) = \{v_1, v_2, v_{(\sum_{i=1}^{c+1} d_i)-3c+3}, \ldots, v_{(\sum_{i=1}^{c+2} d_i)-3c}\}$.

- Append $d_{c+3} - 1$ new vertices to v_{c+3} such that

$$N(v_{c+3}) = \{v_1, v_{(\sum_{i=1}^{c+2} d_i)-3c+1}, \ldots, v_{(\sum_{i=1}^{c+3} d_i)-3c-1}\}, \text{ etc.}$$

It is easy to see that $v_1 v_2 v_3, \ldots, v_1 v_2 v_{c+2}$ form c triangles (cycles) in $G^*(\pi)$, and that $G^*(\pi)$ is a BFS-graph.

We will show that this graph $G^*(\pi)$ is optimal in $\Gamma(\pi)$ when the degree sequence π satisfies certain conditions. We will call such a graphical degree sequence $\pi = (d_1, d_2, \ldots, d_n)$ *obliging* if:

- $\sum_{i=1}^{n} d_i = 2(n + c - 1)$ with non-negative integer c;

- $d_1 \geq d_2 \geq c + 1$;

- $d_4 = d_5 = \cdots = d_{c+2}$ for $c \geq 1$;

- $d_n = 1$.

The following statement identifies the optimal graph with a given obliging sequence. Its proof will be presented in the rest of this section.

Theorem 3.5.1 *If $\pi = (d_1, d_2, \ldots, d_n)$ is an obliging sequence, then $G^*(\pi)$ is optimal (achieving the minimum ABC index) in $\Gamma(\pi)$. In other words,*

$$ABC(G^*(\pi)) \leq ABC(G)$$

for any graph $G \in \Gamma(\pi)$.

3.5.2   Structural properties of the optimal graphs

First, we note the following observation. The proof is through simple calculus and we leave it as an exercise.

Lemma 3.5.1 *Let $f(x, y) = \sqrt{\frac{x+y-2}{xy}}$ where x, y are positive integers. The following statements hold:*

- $f(x, 1)$ *is strictly increasing with respect to x;*

- $f(x, 2) = \frac{\sqrt{2}}{2}$;

- $f(x, y)$ *is strictly decreasing with respect to x if* $y \geq 3$.

With Lemma 3.5.1, direct computation shows the following.

Lemma 3.5.2 ([71]) *Given* $G \in \Gamma(\pi)$ *with* $uv, xy \in E(G)$ *and* $uy, xv \notin E(G)$, *let*
$$G_1 = G - uv - xy + uy + xv.$$
If $\deg(u) \geq \deg(x)$ *and* $\deg(y) \geq \deg(v)$, *then*
$$ABC(G_1) \leq ABC(G)$$
with the equality if and only if $\deg(u) = \deg(x)$ *or* $\deg(v) = \deg(y)$.

We now establish a series of observations, in the form of technical lemmas, regarding the characteristics of the optimal graph in $\Gamma(\pi)$. Essentially, each lemma builds upon the previous one and more specifically characterizes the optimal graph.

Lemma 3.5.3 *Let* $G \in \Gamma(\pi)$ *be a connected graph with three vertices* u, v, w *in G such that*

- $uv \in E(G)$, $uw \notin E(G)$,

- $\deg(u) \geq \deg(w) > \deg(v)$, *and*

- $\deg(u) > \deg(x)$ *for all* $x \in N(w)$.

Then, G is not optimal.

Proof:

We only need to find a graph $G_1 \in \Gamma(\pi)$ such that $ABC(G_1) < ABC(G)$. Two cases are considered, depending on whether v and w have any common neighbor:

- $N(v) \cap N(w) \neq \emptyset$ or $wv \in E(G)$.

  Since $\deg(w) > \deg(v)$, there exists a vertex $y \in N(w)$ such that $y \notin N(v)$. Consider
  $$G_1 = G + uw + vy - uv - wy,$$
  a connected graph in $\Gamma(\pi)$. Since $\deg(u) > \deg(y)$ and $\deg(w) > \deg(v)$, by Lemma 3.5.2 we have $ABC(G_1) < ABC(G)$.

- $N(v) \cap N(w) = \emptyset$ and $wv \notin E(G)$. Since G is connected, there exists a path $P = u, \ldots, s, w$ from u to w.

  - If $uv \notin E(P)$, let
    $$G_1 = G + uw + vs - uv - ws.$$

    Then, G_1 is connected and $G_1 \in \Gamma(\pi)$. We have $ABC(G_1) < ABC(G)$ by Lemma 3.5.2, as $\deg(u) > \deg(s)$ and $\deg(w) > \deg(v)$.

   – If $uv \in E(P)$, then there exists a vertex $y \in N(w)$ such that $y \neq s$
(since $\deg(w) > \deg(v) \geq 1$). Note that $vy \notin E(G)$, since v and w do
not have common neighbors. Let

$$G_1 = G + uw + vy - uv - wy.$$

Then, G_1 is connected and $G_1 \in \Gamma(\pi)$. We have $ABC(G_1) < ABC(G)$
by Lemma 3.5.2, since $\deg(u) > \deg(y)$ and $\deg(w) > \deg(v)$.

□

Lemma 3.5.4 *For a given graphical degree sequence* $\pi = (d_1, \ldots, d_n)$ *with*
$n \geq 3$, *there exists an optimal graph* $G \in \Gamma(\pi)$ *with* $\deg(v_i) = d_i$ *for* $i =$
$1, \ldots, n$ *such that* $\{v_2, v_3\} \subseteq N(v_1)$.

Proof:

First, we show that the optimal graph can be labeled such that $v_1 v_2 \in E(G)$.
Otherwise, no two adjacent vertices have the largest degrees. Let G be an
optimal graph with $\deg(v_i) = d_i$ for $i = 1, \ldots, n$ such that $v_1 v_2 \notin E(G)$,
$\deg(v_1) \geq \deg(v_2) > \deg(v)$ for any $v \in N(v_1)$, and $\deg(v_1) > \deg(x)$ for any
$x \in N(v_2)$. Then, Lemma 3.5.3 indicates that G is not optimal, a contradiction.

   Now, we may continue with an optimal graph G with $\deg(v_i) = d_i$ for
$i = 1, \ldots, n$ such that $v_1 v_2 \in E(G)$. Suppose (for contradiction) that $v_1 v_3 \notin$
$E(G)$. Similar to before, we may assume that $\deg(v_3) > \deg(v)$ for every
$v \in N(v_1) \setminus \{v_2\}$. Furthermore, we claim that there exists a vertex $u \in N(v_3)$
such that $\deg(u) = \deg(v_1) = d_1$. Otherwise, we have $\deg(v_1) > \deg(x)$ for
every $x \in N(v_3)$, and, by Lemma 3.5.3, we may conclude that G is not optimal.

- If $u - v_2$, then $\deg(v_2) = \deg(v_1) = d_1$ and $v_2 v_3, v_2 v_1 \in E(G)$. The conclusion follows from interchanging the labels of v_1 and v_2.

- If $u \neq v_2$, then $d_1 = \deg(u) \leq \deg(v_3) \leq d_2 \leq d_1$, implying that $\deg(u) =$
$\deg(v_1) = \deg(v_2) = \deg(v_3) = d_1$. Now that all these four vertices share
the common largest degree, through relabeling vertices we may assume that
the shortest path $P = v_1, x, \ldots, v_3$ from v_1 to v_3 (since G is connected) does
not contain v_2. Since $v_1 \in N(x) \setminus N(v_3)$ and $\deg(v_3) > \deg(x)$, there must
exist a vertex $y \in N(v_3) \setminus N(x)$ that is not on P. Then

$$G' = G + v_1 v_3 + xy - v_1 x - v_3 y \in \Gamma(\pi)$$

and $ABC(G') \leq ABC(G)$ by Lemma 3.5.2. Hence, G' is an optimal graph
with $v_1 v_2, v_1 v_3 \in E(G')$.

□

   Next, one can show that in an optimal graph of $\Gamma(\pi)$ (when π is obliging),
three of the vertices of the largest degrees must induce a triangle.

Lemma 3.5.5 *Let* $\pi = (d_1, d_2, \ldots, d_n)$ *be an obliging graphical degree sequence with* $c > 0$. *Then there exists an optimal graph* $G \in \Gamma(\pi)$ *with* $\deg(v_i) = d_i$ *for* $i = 1, \ldots, n$ *such that* v_1, v_2, v_3 *form a triangle.*

Proof:

As the first step, the following claims that v_1, a vertex of the largest degree, belongs to a cycle.

Claim 1. There exists an optimal graph $G \in \Gamma(\pi)$ such that $\{v_2, v_3\} \subseteq N(v_1)$ and there exists a cycle C_{t_1} in G such that $v_1 \in V(C_{t_1})$.

Proof of Claim 1:

Otherwise, Claim 1 does not hold for any optimal graph $G \in \Gamma(\pi)$.

Recall that Lemma 3.5.4 states that there exists an optimal graph $G \in \Gamma(\pi)$ such that $\{v_2, v_3\} \subseteq N(v_1)$. Then, v_1 is simply not on any cycle of G. Since $d_1 \geq c + 1$, there exists a shortest path $P = v, \ldots, v_1, \ldots, x, y$ from a vertex v in a cycle C_{t_1} to a pendant vertex y through v_1. Let u and w be two adjacent vertices (that are different from v) in $V(C_{t_1})$. We consider two cases:

- If $\deg(w) \leq \deg(x)$, let

$$G_1 = G + ux + wy - wu - xy.$$

By Lemma 3.5.2, $ABC(G_1) \leq ABC(G)$. Now, $G_1 \in \Gamma(\pi)$ is also optimal and v_1 is in some cycle of G_1, a contradiction.

- If $\deg(u) \leq \deg(x)$, let

$$G_2 = G + uy + wx - wu - xy.$$

Through similar discussion as above, we see that $G_2 \in \Gamma(\pi)$ is also optimal and v_1 is in some cycle of G_2, a contradiction.

Consequently, we must have $\min\{\deg(u), \deg(w)\} > \deg(x)$. Let z be the neighbor of x on P other than y, and consider the graphs

$$G_1' = G + uz + wx - wu - xz,$$

and

$$G_2' = G + ux + wz - wu - xz.$$

Exactly the same reasoning shows that $\min\{\deg(u), \deg(w)\} > \deg(z)$. This process can be repeated to show that $\min\{\deg(u), \deg(w)\} > \deg(v_1)$, which is a contradiction. $\qquad\square$

Next we show that some vertex of second largest degree is also on the same cycle.

Claim 2. There exists an optimal graph $G \in \Gamma(\pi)$ such that there is a cycle C_{t_1} containing $v_1 v_2 \in E(C_{t_1})$ and $v_3 \in N(v_1)$.

Proof of Claim 2:

Otherwise, Claim 2 does not hold for any optimal graph $G \in \Gamma(\pi)$. Since Claim 1 guarantees the existence of an optimal graph $G \in \Gamma(\pi)$ such that $v_1 \in V(C_{t_1})$ and $\{v_2, v_3\} \subseteq N(v_1)$, it must be the case that $v_2 \notin V(C_{t_1})$. Consider two further cases:

Case 1: v_2 is contained in a path from v_1 to a leaf z, say $P = v_1, v_2, x, y, \ldots, z$. Further, assume that this is the shortest such path. Choose $\{u, v\} \subseteq V(C_{t_1})$ such that $uv \in E(C_{t_1})$ and suppose $\max\{\deg(u), \deg(v)\} = \deg(u)$.

Following the same reasoning as before, if $\deg(u) \geq \deg(x)$ and

$$G_1 = G + uv_2 + vx - uv - v_2 x,$$

Lemma 3.5.2 implies $ABC(G_1) \leq ABC(G)$. Then, $G_1 \in \Gamma(\pi)$ is optimal with $v_1 v_2$ on a cycle, a contradiction.

Thus, $\max\{\deg(u), \deg(v)\} < \deg(x)$. The same argument applies to y, etc. In the end we have $\max\{\deg(u), \deg(v)\} < \deg(z) = 1$, a contradiction.

Case 2: v_2 is not contained in any path from v_1 to a leaf. Then, $G - v_1 v_2$ has two components and any leaf must be in the same component as v_1 (for otherwise the path from v_1 to this leaf will have to contain v_2).

Then, the component containing v_2 must be cyclic, implying that $c \geq 2$. We may consider a cycle C_{t_2} in this component, together with a shortest path $P' = v_2, v_1, x', y', \ldots, z$ from v_2 to a leaf z that contains v_1. We may now repeat the same argument as that in Case 1 to get a contradiction. □

As expected, next we show that v_3, the vertex with the third largest degree, is also on the same cycle.

Claim 3. There is an optimal graph $G \in \Gamma(\pi)$ such that $\{v_1 v_2, v_1 v_3\} \subseteq E(C_{t_1})$.

Proof of Claim 3:

We already know from Claim 2 that there is an optimal graph $G \in \Gamma(\pi)$ such that $v_1 v_2 \in E(C_{t_1})$ and $v_3 \in N(v_1)$. If $v_3 \notin V(C_{t_1})$, then $v_2 v_3 \notin E(G)$ (for otherwise v_1, v_2, v_3 forms the obvious triangle).

Now, let u be a neighbor of v_2 on C_{t_1} other than v_1, and let v be a neighbor of v_3 other than v_1. Then, $uv \notin E(G)$, $\deg(v_2) \geq \deg(v)$, and $\deg(v_3) \geq \deg(u)$. Let

$$G_1 = G + v_2 v_3 + uv - vv_3 - uv_2,$$

Lemma 3.5.2 implies that $G_1 \in \Gamma(\pi)$ is optimal with $v_1 v_2$ and $v_1 v_3$ lying on a cycle. □

Lastly, one can claim that this common cycle of v_1, v_2, v_3 is indeed a triangle.

Claim 4. There is an optimal graph $G \in \Gamma(\pi)$ such that v_1, v_2, v_3 form a triangle.

Proof of Claim 4:

Let us only consider the non-trivial case that $d_2 \geq 3$.

From Claim 3, we have an optimal graph $G \in \Gamma(\pi)$ such that $\{v_1v_2, v_1v_3\} \subseteq E(C_{t_1})$. For contradiction, assume that $v_2v_3 \notin E(G)$.

Similar to the previous proof, let u be a neighbor of v_2 not on C_{t_1} and v a neighbor of v_3 on C_{t_1} other than v_1. Now let

$$G_1 = G + v_2v_3 + uv - vv_3 - uv_2.$$

Then Lemma 3.5.2 applies as before and $G_1 \in \Gamma(\pi)$ is optimal with the triangle $v_1v_2v_3$. □

Since Claim 4 is exactly the statement of the lemma, this concludes our proof. □

Now that we have established the existence of one triangle (formed by the vertices of the largest degrees) in the optimal graph, following exactly the same arguments, one can show that there is an edge between v_1 and v_i, the vertex of the next largest degree that is not on any cycle yet, etc. Eventually one concludes that there are exactly c triangles, as stated in the following. We leave the technical details as exercises.

Lemma 3.5.6 *Let π be an obliging graphical sequence. There exists an optimal graph $G \in \Gamma(\pi)$ such that $v_1v_2v_3, \ldots, v_1v_2v_{c+2}$ form c triangles.*

3.5.3   Proof of Theorem 3.5.1

From the previous section, we have c triangles $v_1v_2v_3, \ldots, v_1v_2v_{c+2}$ in an optimal graph $G \in \Gamma(\pi)$. We now create an ordering $\prec$ of $V(G)$ following the breadth-first search:

- Let $v_1 \prec v_2 \prec \cdots \prec v_{c+2}$;

- Add all neighbors $u_{c+3}, \ldots, u_{d_1+1}$ of $N(v_1)\backslash\{v_2, \ldots, v_{c+2}\}$ to the ordered list such that $u \prec v$ whenever $\deg(u) > \deg(v)$ (the ordering between some vertices can be arbitrary);

- Add all neighbors $u_{d_1+2}, u_{d_1+3}, \ldots, u_{d_1+d_2-c}$ of $N(v_2)\backslash\{v_1, v_3, \ldots, v_{c+2}\}$ to the ordered list such that $u \prec v$ whenever $\deg(u) > \deg(v)$ (the ordering between some vertices can be arbitrary); Similarly, we can add the vertices of $N(v_3)\backslash\{v_1, v_2\}, \ldots, N(v_{c+2})\backslash\{v_1, v_2\}$ to the ordered list;

- Add the vertices in $N(x)\backslash\{v_1\}$ to the ordered list, where $\deg(x) = \max\{\deg(y) : y \in N(v_1)\backslash\{v_2, v_3, \ldots, v_{c+2}\}\}$;

- Repeat the last step until all vertices are added to the list.

To prove Theorem 3.5.1, it is now sufficient to verify the conditions in Definition 3.5.1 for such an ordering $v_1 \prec v_2 \prec \cdots \prec v_n$.

First, note that conditions (1) and (3) are obvious.

From our construction, $u \prec v$ implies $h(u) \leq h(v)$. Let $V_i(G)$ denote the set of vertices of height i in G, with $V_0(G) = \{v_1\}$. For $v \in V_i(G)$, $i > 0$, we call the unique vertex $u \in N(v) \cap V_{i-1}(G)$ the parent of v. Hence $u \prec v$ if u is the parent of v. Moreover, because the vertices are appended to the ordered list recursively, if there are two edges $uu_1 \in E(G)$ and $vv_1 \in E(G)$ such that $u \prec v$, $h(u) = h(u_1) + 1$ and $h(v) = h(v_1) + 1$, then $u_1 \prec v_1$.

Lastly, we will show that $\deg(u) \geq \deg(v)$ holds for any $u, v \in V(G)$ with $u \prec v$. Suppose otherwise, let v_i be the first vertex where this fails. That is, $v_i \prec u$ in the ordering and $\deg(v_i) < \deg(u)$ for some $u \in V(G)$. From the construction, we know that $v_i \notin \{v_1, v_2, v_3, \ldots, v_{c+2}\}$, and if $v \prec v_i$ then $\deg(v) \geq \deg(u)$ holds for each u with $v \prec u$.

Consider, now, the vertex of the largest degree in the remaining vertices and let v_j be the first such vertex in the ordering. In other words, j is the smallest integer such that $v_i \prec v_j$ and $\deg(v_j) = \max\{\deg(v_t) : i+1 \leq t \leq n\}$.

Recall that $v_i \prec v_j$ and $\deg(v_i) < \deg(v_j)$ by our assumption. Let w_i and w_j be the parents of v_i and v_j, respectively. Since $\deg(v_i) < \deg(v_j)$, we must have $w_i \neq w_j$ and $w_i \prec w_j$. Note that $w_i v_j \notin E(G)$. Otherwise, there is a cycle in G containing w_i, w_j and v_j in addition to the c triangles.

- If $w_i v_i$ is on the shortest path that connects w_j and v_1, then

$$w_i \prec v_i \prec w_j \prec v_j$$

  and

$$\deg(w_i) \geq \deg(v_j) > \deg(w_j)$$

  by our choices of v_i and v_j.

  We claim that there exists some $y \in N(v_j) \backslash \{w_j\}$ such that $\deg(w_i) = \deg(v_j) = \deg(y)$ and $v_i y \notin E(G)$. The proof is rather similar to the arguments that have been repeated so far and we leave it as an exercise.

  Now let

$$G_1 = G + w_i v_j + v_i y - w_i v_i - v_j y,$$

  Lemma 3.5.2 implies that G_1 is optimal.

- If $w_i v_i$ is not on the shortest path that connects w_j and v_1, then similar to before we must have $w_j v_i \notin E(G)$.

  Let

$$G_1 = G + w_i v_j + w_j v_i - w_i v_i - w_j v_j \in \Gamma(\pi),$$

  again Lemma 3.5.2 applies to show that $ABC(G_1) \leq ABC(G)$.

In both cases a new graph G_1 (with the same c triangles) was obtained such that $G_1 \in \Gamma(\pi)$ and $ABC(G_1) \leq ABC(G)$. Consequently, we will replace

v_i with v_j in the ordering, then add the remaining vertices according to the same rule.

It is easy to see that this process terminates after at most $n - c - 2$ steps (every time a new ordering is introduced with a new graph with the same c triangles and at most as large $ABC(\cdot)$), resulting in an optimal graph $G_t \in \Gamma(\pi)$ such that $\deg(u) \geq \deg(v)$ holds for every two vertices $u, v \in V(G)$ and $u \prec v$. This concludes the proof of Theorem 3.5.1.

3.5.4 Acyclic, unicyclic, and bicyclic optimal graphs

When Theorem 3.5.1 is applied to acyclic, unicyclic, and bicyclic graphs, corresponding optimal graphs with a given degree sequence can be easily identified.

For this purpose, we let

$$\mathcal{U}(\pi) = \{G \mid G \text{ is a unicylic graph with degree sequence } \pi\}$$

and

$$\mathcal{B}(\pi) = \{G \mid G \text{ is a bicyclic graph with degree sequence } \pi\}.$$

In particular, we denote the graph $G^*(\pi)$ by $U^*(\pi)$ if $c = 1$ and $B^*(\pi)$ if $c = 2$.

In the case that π is the degree sequence of a tree, applying Theorem 3.5.1 yields the same extremal results that have been discussed earlier. For unicyclic graphs, we have the following. We leave the simple proof as an exercise.

Theorem 3.5.2 *Among all unicyclic graphs with degree sequence π and $d_n = 1$, $U^*(\pi)$ achieves the minimum ABC index.*

The situation for bicyclic graphs is much more complicated. We start with some more definitions:

- Given $p + q - 1 = n$, denote by $B(p, q)$ the bicyclic graph of order n obtained from two vertex-disjoint cycles C_p and C_q by identifying vertices u of C_p and v of C_q.

- Given $p + q + r - 1 = n$, denote by $B(p, r, q)$ the bicyclic graph of order n obtained from two vertex-disjoint cycles C_p and C_q by joining vertices u of C_p and v of C_q by a new path $u, u_1, u_2, \ldots, u_{r-1}, v$ of length $r \geq 1$.

- Given $k + l + m - 1 = n$, denote by $B(P_k, P_l, P_m)$ ($1 \leq m \leq \min\{k, l\}$) a bicyclic graph of order n obtained from three pairwise internal disjoint paths $P_k = x, v_1, v_2, \ldots, v_{k-1}, y$, $P_l = x, u_1, u_2, \ldots, u_{l-1}, y$, and $P_m = x, w_1, w_2, \ldots, w_{m-1}, y$.

- Given $p + q + p_1 + \cdots + p_s - 1 = n$, denote by $B(p, q; p_1, p_2, \ldots, p_s)$ a bicyclic graph of order n obtained from $B(p, q)$ by appending s paths on the common vertex of the two cycles. Here, s is the number of leaves and $p_1, p_2, \ldots, p_s$ denote the lengths of the s paths, respectively.

Theorem 3.5.3 *For bicyclic graphs of degree sequence* $\pi = (d_1, d_2, \ldots, d_n)$ *with* s *leaves, the following holds:*

- *If* $d_n = 2$ *and* $d_2 \geq 3$, *then for every* $G \in \mathcal{B}(\pi)$,

$$ABC(G) \geq \frac{\sqrt{2}}{2}n + \frac{2}{3}$$

  with equality if and only if G *is* $B(p, 1, q)$ *or* $B(P_k, P_l, P_1)$ *with* $p + q = k + l = n$.

- *If* $d_n = 2$ *and* $d_2 = 2$, *then every graph in* $\mathcal{B}(\pi)$ *is isomorphic to some* $B(p, q)$ *with* $p + q = n$, *and* $ABC(G) = \frac{\sqrt{2}}{2}(n+1)$.

- *If* $d_n = 1$, $d_2 = 2$ *and* $s \leq \frac{n-5}{2}$, *then for every* $G \in \mathcal{B}(\pi)$,

$$ABC(G) \geq \frac{\sqrt{2}}{2}(n+1)$$

  with equality if and only if G *is isomorphic to some* $B(p, q; p_1, p_2, \ldots, p_s)$ *with* $p_i \geq 2$ *for* $1 \leq i \leq s$.

- *If* $d_n = 1$ *and* $d_2 = 2$ *and* $s > \frac{n-5}{2}$, *then for every* $G \in \mathcal{B}(\pi)$,

$$ABC(G) \geq (n - s - 5)\sqrt{2} + (2s - n + 5)\sqrt{\frac{s+3}{s+4}}$$

  with equality if and only if G *is isomorphic to* $B(3, 3; 2, \ldots, 2, 1, \ldots, 1)$ *with*

$$p_1 = \cdots = p_{n-s-5} = 2 \text{ and } p_{n-s-4} = \cdots = p_s = 1.$$

- *If* $d_n = 1$ *and* $d_2 \geq 3$, *then* $B^*(\pi)$ *is an optimal graph in the set* $\mathcal{B}(\pi)$.

Proof:

The extremality immediately follows from Theorem 3.5.1. We briefly discuss the bounds for each case and leave the details as exercises.

(1) If $d_n = 2$ and $d_2 \geq 3$, then the only possible degree sequence is $\pi = (3, 3, 2^{(n-2)})$ and G is $B(p, r, q)$ or $B(P_k, P_l, P_m)$. It is easy to see that

$$ABC(B(p, 1, q)) = ABC(B(P_k, P_l, P_1)) = \frac{\sqrt{2}}{2}n + \frac{2}{3}$$

$$< ABC(B(p, r, q)) = ABC(B(P_k, P_l, P_m)) = \frac{n+1}{2}\sqrt{2}$$

for $r, m > 1$.

(2) If $d_n = 2$ and $d_2 = 2$, then G is $B(p, q)$ with $p + q - 1 = n$. It is easy to see that

$$ABC(B(p, q)) = \frac{n+1}{2}\sqrt{2}.$$

(3) If $\pi = (d_1, 2^{(n-s-1)}, 1^{(s)})$ and $s \leq \frac{n-5}{2}$, then all possible graphs are of the form $B(p, q; p_1, p_2, \ldots, p_s)$. All possible pairs of degrees of adjacent vertices that occur are $(d_1, 1)$, $(d_1, 2)$, $(2, 2)$, $(d_1, 2)$, and $(2, 1)$. Using Lemma 3.5.1, it is easy to see that the minimum ABC index of $\frac{n+1}{2}\sqrt{2}$ is obtained whenever $p_i \geq 2$ for $1 \leq i \leq s$.

(4) If $\pi = (d_1, 2^{(n-s-1)}, 1^{(s)})$ and $s > \frac{n-5}{2}$, one can still argue as in the previous case. By Lemma 3.5.1, the unique optimal graph of this case is $B(3, 3; 2, \ldots, 2, 1, \ldots, 1)$ with ABC index

$$(2s + 5 - n)\sqrt{\frac{s+3}{s+4}} + (n - s - 2)\sqrt{2}.$$

(5) Directly from Theorem 3.5.1. $\qquad\qquad\qquad\qquad\qquad\qquad \square$

3.6 Graphs with a given matching number

Before ending this chapter, we utilize Lemma 1.7.1 from the introduction and show its application in extremal problems with respect to a number of degree-based indices in graphs with a given matching number. Most of the results here were introduced in [15] and the references thereof.

Similar analysis can be done on some distance-based indices following similar logic. We leave those to interested readers.

Through this discussion, we hope to shed some light on general approaches to deal with degree-based (as well as distance-based and other) indices in graphs with given parameters using previously established facts from theoretical studies of graphs.

3.6.1 Generalized Randić index

First, note that for $\alpha > 0$ the *generalized Randić index*

$$R_\alpha(G) = \sum_{uv \subseteq E(G)} (\deg(u) \deg(v))^\alpha$$

satisfies the conditions of Lemma 1.7.1. Applying this lemma, together with further discussion, leads to the following.

Theorem 3.6.1 *Among connected graphs of order n and matching number β, $R_\alpha(\cdot)$ (for some $\alpha > 1$) is maximized by*

$$\widehat{G} = K_s + (K_{2\beta - 2s + 1} \cup \overline{K_{n+s-2\beta-1}})$$

for some s.

Proof:

Let $\widehat{G}$ be the extremal graph under consideration. Lemma 1.7.1 implies that

$$\widehat{G} = K_s + (K_{n_1} \cup K_{n_2} \cup \cdots \cup K_{n_t})$$

for some integer s and odd integers $n_1, n_2, \ldots, n_t$. Then, direct computation yields

$$R_\alpha(\widehat{G}) = \sum_{i=1}^{t} \frac{n_i(n_i - 1)(n_i + s - 1)^{2\alpha}}{2}$$

$$+ \sum_{i=1}^{t} s n_i [(n-1)(n_i + s - 1)]^\alpha$$

$$+ \frac{(s^2 - s)(n-1)^{2\alpha}}{2}.$$

Now, it suffices to show that the above expression attains its maximum if and only if we have

$$n_1 = n_2 = \cdots = n_{t-1} = 1 \text{ and } n_t = 2\beta - 2s + 1,$$

or some permutation thereof.

Supposing, without loss of generality, that $1 \leq n_1 \leq n_2 \leq \cdots \leq n_t$, and for contradiction that $3 \leq n_i \leq n_j$, consider

$$G' = K_s + (K_{n_1} \cup \cdots \cup K_{n_i - 2} \cup \cdots \cup K_{n_j + 2} \cup \cdots \cup K_{n_t}).$$

Then,

$R_\alpha(G') - R_\alpha(\widehat{G})$

$$= \frac{(n_i - 2)(n_i - 3)(n_i + s - 3)^{2\alpha}}{2} + s(n_i - 2)[(n-1)(n_i + s - 3)]^\alpha$$

$$+ \frac{(n_j + 2)(n_j + 1)(n_j + s + 1)^{2\alpha}}{2} + s(n_j + 2)[(n-1)(n_j + s + 1)]^\alpha$$

$$- \frac{n_i(n_i - 1)(n_i + s - 1)^{2\alpha}}{2} - s n_i [(n-1)(n_i + s - 1)]^\alpha$$

$$- \frac{n_j(n_j - 1)(n_j + s - 1)^{2\alpha}}{2} - s n_j [(n-1)(n_j + s - 1)]^\alpha$$

$$= \frac{n_j^2 - n_j}{2}[(n_j + s + 1)^\alpha + (n_j + s - 1)^\alpha][(n_j + s + 1)^\alpha - (n_j + s - 1)^\alpha]$$

$$- \frac{n_i^2 - n_i}{2}[(n_i + s - 3)^\alpha + (n_i + s - 1)^\alpha][(n_i + s - 1)^\alpha - (n_i + s - 3)^\alpha]$$

$$+ s n_j (n-1)^\alpha [(n_j + s + 1)^\alpha - (n_j + s - 1)^\alpha]$$

$$- s n_i (n-1)^\alpha [(n_i + s - 1)^\alpha - (n_i + s - 3)^\alpha]$$

$$+ (2n_j + 1)(n_j + s + 1)^{2\alpha} - (2n_i - 3)(n_i + s - 3)^{2\alpha}$$

$$+ 2s(n-1)^\alpha (n_j + s + 1)^\alpha - 2s(n-1)^\alpha (n_i + s - 3)^\alpha$$

$$> 0$$

since $(n_j + s + 1)^\alpha - (n_j + s - 1)^\alpha > (n_i + s - 1)^\alpha - (n_i + s - 3)^\alpha$ for $\alpha > 1$, contradicting the extremality of $\widehat{G}$.

$\square$

3.6.2 Zagreb indices based on edge degrees

First, recall the definitions of the first and second Zagreb indices as

$$M_1(G) = \sum_{u \in V(G)} \deg^2(u) \text{ and } M_2(G) = \sum_{uv \in E(G)} \deg(u)\deg(v).$$

Miličević, Nikolić and Trinajstić reformulated the Zagreb indices by using edge-degrees instead of vertex-degrees, as

$$EM_1(G) = \sum_{e \in E(G)} \deg^2(e) \text{ and } EM_2(G) = \sum_{e \sim f} \deg(e)\deg(f),$$

where $\deg(e)$ denotes the degree of the edge e in G, defined as $\deg(e) = \deg(u) + \deg(v) - 2$ with $e = uv$; and $e \sim f$ the adjacency of e and f (i.e., they share a common vertex in G). The first and second *reformulated variable Zagreb indices* $EM_1^\alpha(G)$ and $EM_2^\alpha(G)$ are defined as

$$EM_1^\alpha(G) = \sum_{e \in E(G)} \deg^{2\alpha}(e) \text{ and } EM_2^\alpha(G) = \sum_{e \sim f} (\deg(e)\deg(f))^\alpha$$

for some $\alpha \neq 0$. When $\alpha = 1$, they are the first and second reformulated Zagreb indices. In this section, we consider the *first reformulated variable Zagreb index* $EM_1^\alpha(G)$.

Recall that $M_2(\cdot)$ is simply a special case of $R_\alpha(\cdot)$, and it is not hard to notice the similarity between the reformulated variable Zagreb indices and $R_\alpha(\cdot)$. Then, it is natural to expect similar results on such edge-degree-based indices.

First, we introduce the following simple observation, which is very similar to that of other indices. The proof is left as an exercise.

Proposition 3.6.1 *Let G be a connected graph with at least three vertices and not complete. Then*

$$EM_1^\alpha(G) < EM_1^\alpha(G + e)$$

for any $e \notin E(G)$ and $\alpha > 0$.

In the case of $\alpha > 0$, applying Lemma 1.7.1 leads to the following. The idea is very similar to the above proof for $R_\alpha(\cdot)$ and we leave some technical computations as exercises.

Theorem 3.6.2 *Among connected graphs of order n and matching number β, $EM_1^\alpha(G)$ (for some $\alpha > 0$) is maximized by*

$$\widehat{G} = K_s + (K_{2\beta-2s+1} \cup \overline{K_{n+s-2\beta-1}})$$

for some s.

Proof:

For such an extremal graph $\widehat{G}$, Lemma 1.7.1 implies that

$$\widehat{G} = K_s + (K_{n_1} \cup K_{n_2} \cup \cdots \cup K_{n_t}).$$

Then,

$$EM_1^\alpha(\widehat{G}) = \sum_{i=1}^{t} \frac{n_i(n_i-1)[2(n_i+s-2)]^{2\alpha}}{2}$$

$$+ \sum_{i=1}^{t} sn_i(n_i+n+s-4)^{2\alpha}$$

$$+ \frac{(s^2-s)(2n-4)^{2\alpha}}{2}.$$

Similar analysis of the above formula as before shows that it attains its maximum if and only if $n_1 = n_2 = \cdots = n_{t-1} = 1$ and $n_t = 2\beta - 2s + 1$ (and hence the conclusion). $\qquad\square$

Depending on the specific value of the matching number β (compared with n), explicit formulas for the above upper bounds can be obtained through rather tedious algebra. We list the statement here without a proof.

Theorem 3.6.3 *Let G be a connected graph with $n \geq 4$ vertices and matching number β, $5 \leq \beta \leq \lfloor \frac{n}{2} \rfloor$. Further, let r be the largest root of the cubic equation*

$$33x^3 + (n-77)x^2 + (17n+32-3n^2)x - n^3 + 5n^2 - 12n = 0.$$

Then, we have

- *if $\beta = \lfloor \frac{n}{2} \rfloor$, then*

$$EM_1(G) \leq \frac{n(n-1)}{2}$$

  with equality if and only if $G \cong K_n$;

- *if $r < \beta \leq \lfloor \frac{n}{2} \rfloor - 1$, then*

$$EM_1(G) \leq 32\beta^4 - 104\beta^3 + (108+8n)\beta^2 - (40+12n)\beta + n^3 - 5n^2 + 12n$$

  with equality if and only if

$$G \cong K_1 + (K_{2\beta-1} \cup \overline{K_{n-2\beta}});$$

- if $5 \le \beta < r$, then

$$EM_1(G) \le -\beta^4 + (6 - n)\beta^3 + (3n^2 - 8n - 1)\beta^2 + (n^3 - 8 + 17n - 8n^2)\beta$$

with equality if and only if

$$G \cong K_\beta + \overline{K_{n-\beta}};$$

- if $\beta = r$, then

$$EM_1(G) \le 32r^4 - 104r^3 + (108 + 8n)r^2 - (40 + 12n)r + n^3 - 5n^2 + 12n$$
$$= -r^4 + (6 - n)r^3 + (3n^2 - 8n - 1)r^2 + (n^3 - 8 + 17n - 8n^2)r$$

with equality if and only if

$$G \cong K_1 + (K_{2\beta-1} \cup \overline{K_{n-2\beta}}) \text{ or } G \cong K_\beta + \overline{K_{n-\beta}}.$$

3.6.3   The Atom-bond connectivity index

Recall that adding an edge to a non-complete graph strictly increases its ABC index [21]. We now have the following immediate consequence.

Theorem 3.6.4 *The maximum ABC index among all connected graphs of order n and matching number β is achieved by a graph of the form*

$$\hat{G} = K_s + (K_{n_1} \cup K_{n_2} \cup \cdots \cup K_{n_t})$$

for some s and t with $s + n_1 + \ldots + n_t = n$.

Exercises

1. Prove Proposition 3.1.1.

2. Complete the first part of the proof of Proposition 3.1.2 by showing the case for $i = 1$.

3. Complete the second part of the proof of Proposition 3.1.2 by verifying the expression of $R_1(T') - R_1(T)$.

4. Prove that among trees of a given order, $R_\alpha(\cdot)$ is minimized by the path for positive α and minimized by the star for negative α.

5. Prove the following identity for all graphs G:

$$\sum_{uv \in E(G)} (\deg(u) + \deg(v)) = M_1(G) = \sum_{v \in V(G)} \deg(v)^2.$$

6. Prove the following inequality for the first Zagreb index of a graph G with n vertices and m edges:

$$M_1(G) \geq \frac{(2m)^2}{n}.$$

7. Prove that the process of "switching" or "reversing" described in Remark 3.2.1 will terminate after finitely many steps.

8. Prove that, with a given degree sequence, if condition (3.3) is satisfied by every path of a tree T, then T is a greedy tree.

9. Prove Lemma 3.2.2.

10. Prove that a tree (of a given degree sequence) satisfying Lemma 3.2.2 is an alternating greedy tree.

11. Prove Theorem 3.3.2.

12. Prove Theorem 3.3.4.

13. First, prove Corollaries 3.3.1 to 3.3.5. Then, state and prove the analogous statements for de-escalating functions.

14. First, prove Proposition 3.4.2. Then, use it to prove Proposition 3.4.3.

15. Prove a stronger version of Lemma 3.4.1, that any extremal tree that maximizes $M_2(\cdot) - M_1(\cdot)$ cannot have diameter more than 4.

16. Fill in the details of the proof of Theorem 3.4.1.

17. Prove Proposition 3.4.6.

18. Prove Lemma 3.5.1. Then, use it to show Lemma 3.5.2.

19. Prove Lemma 3.5.6.

20. Prove Theorem 3.5.2.

21. Fill in the details of the proof of Theorem 3.6.2.

4

Independent sets: Merrifield-Simmons index and Hosoya index

4.1 History and terminologies

This chapter is devoted to two topological indices of a very similar nature: the Merrifield-Simmons index and the Hosoya index. They are two important graph invariants based on counting subsets: in the case of the Merrifield-Simmons index, independent subsets of vertices; in the case of the Hosoya index, independent subsets of edges (matchings).

Definition 4.1.1 *The* Merrifield-Simmons index *of a graph G, henceforth denoted by $\sigma(G)$, is the total number of independent sets of vertices of G. The* Hosoya index *of a graph G, henceforth denoted by $Z(G)$, is the total number of matchings of G.*

Let us consider an example to illustrate the definition. The graph G in Figure 4.1 has six independent sets: the empty set, four single-vertex sets, and one independent set of two vertices. Hence $\sigma(G) = 6$. Likewise, it has eight matchings: the empty set, five single-edge sets, and two matchings consisting of two edges. Hence $Z(G) = 8$.

The quantity $Z(G)$ associated with a graph was introduced to the chemical literature in 1971 by the Japanese chemist Haruo Hosoya [49]. This was also the first time that the term topological index was used. Originally, it only referred to what is now known as the Hosoya index, but the expression is now widely used for other graph invariants in the context of mathematical

FIGURE 4.1
Example graph for the Merrifield-Simmons index and the Hosoya index.

chemistry. It was found that the quantity $Z(G)$ of the molecular graph of a saturated hydrocarbon correlated well with a variety of physico-chemical properties, such as boiling points, see [50,51]. It also plays a role in the theory of conjugated π-electron systems, as does the graph energy, which we will consider in the following chapter, and which is closely related to the Hosoya index.

The Merrifield-Simmons index was introduced by chemists Merrifield and Simmons in a series of papers [78–81] and a book [77] as part of an elaborate theory involving finite-set topology. The number of independent sets occurred in this framework as the number of open sets of a certain finite topology, and of all the aspects of their theory, it probably received the most attention. It is interesting to see that the mathematical study of the number of independent sets of a graph was initiated around the same time in [89], where the name Fibonacci number of a graph was coined for the quantity $\sigma(G)$. After some early results had been obtained in the 1980s, there was a bit of a break in the mathematical study of Merrifield-Simmons index and Hosoya index until they saw a sudden rise in popularity in the 2000s, when a lot of articles were published on properties of these two invariants, with a focus mostly on extremal problems. A selection of important results will be presented in this chapter. The interested reader is referred to the surveys [106] and [105] for a more in-detail account of the literature on the Merrifield-Simmons index and the Hosoya index.

4.2   Merrifield-Simmons index and Hosoya index: elementary properties

Both the Merrifield-Simmons index and the Hosoya index satisfy basic monotonicity properties, as stated in the following lemma (cf. Section 1.7):

Lemma 4.2.1 *If edges are removed from a graph, then the Merrifield-Simmons index increases, while the Hosoya index decreases.*

If vertices are removed from a graph, then the Merrifield-Simmons index decreases. The Hosoya index does not increase, and decreases strictly if at least one of the vertices that are removed is not an isolated vertex.

Proof:

We only consider the situation that edges are removed, the other statement follows analogously. It is clear that all independent sets stay independent, and at least one new independent set (consisting of the two endvertices) is added for every edge that is removed. Hence the Merrifield-Simmons index increases. On the other hand, matchings in the graph resulting from the removal of edges are also matchings in the original graph, and removing an edge also always means that at least one matching (consisting only of the single edge) is removed. Thus the Hosoya index must decrease. □

The following properties are very useful in computing the Merrifield-Simmons index and the Hosoya index, and they will also be used frequently in proofs throughout this chapter.

Lemma 4.2.2 • *If $G_1, G_2, \ldots, G_k$ are the connected components of a graph G, then we have*

$$\sigma(G) = \prod_{j=1}^{k} \sigma(G_j).$$

• *For every vertex v of G, we have*

$$\sigma(G) = \sigma(G - v) + \sigma(G - N[v]).$$

• *For every edge e of G whose ends are v and w, we have*

$$\sigma(G) = \sigma(G \quad e) \quad \sigma(G - (N[v] \cup N[w])).$$

Proof:

For the first statement, we simply note that every independent set of G induces independent sets in all components, and conversely, the union of independent sets in all components is always an independent set of G. The second formula follows from the fact that the set of independent vertex sets of G can be decomposed into two parts: those that do not contain v (all independent subsets of $G - v$), and those that do (obtained by adding v to an arbitrary independent subset of $G - N[v]$ – note that none of v's neighbors can be contained). Finally, the third statement follows from a similar argument: all independent vertex subsets of G are also independent sets in $G - e$, and vice versa, except for those that contain both ends of e. These sets are obtained by adding v and w to an arbitrary independent set of $G - (N[v] \cup N[w])$. □

Lemma 4.2.3 • *If $G_1, G_2, \ldots, G_k$ are the connected components of a graph G, then we have*

$$Z(G) = \prod_{j=1}^{k} Z(G_j).$$

• *For every vertex v of G, we have*

$$Z(G) = Z(G - v) + \sum_{w \in N(v)} Z(G - \{v, w\}).$$

• *For every edge e of G whose ends are v and w, we have*

$$Z(G) = Z(G - e) + Z(G - \{v, w\}).$$

Proof:

The proof of the first formula is analogous to Lemma 4.2.2. To obtain the second statement, we have to distinguish two cases again: a matching either does not contain any edge incident with v, or exactly one of them. In the former case, a matching of $G - v$ remains. In the latter case, the matching consists of an edge vw (for some neighbor w of v) and an arbitrary matching of $G - \{v, w\}$. Lastly, for every edge $e = vw$, we can distinguish between matchings that do not contain e (which are matchings of $G - e$) and matchings that contain it (obtained by adding e to an arbitrary matching of $G - \{v, w\}$).   □

The important special case of a pendant vertex (a vertex of degree 1) is specifically considered in the following corollary:

Corollary 4.2.1 *If v is a pendant vertex in a graph G and w its unique neighbor, then we have*

$$\sigma(G) = \sigma(G - v) + \sigma(G - \{v, w\})$$

and

$$Z(G) = \sigma(G - v) + Z(G - \{v, w\}).$$

The Merrifield-Simmons index and the Hosoya index do not only have very similar definitions, there is also a direct connection between the two: recall that the vertices of the line graph $\mathcal{L}(G)$ of a graph G are the edges of G, and two vertices in $\mathcal{L}(G)$ are connected by an edge if and only if the corresponding edges in G have a common vertex. It is easy to see that matchings in G correspond precisely to independent sets in $\mathcal{L}(G)$. Therefore, the following is immediate:

Proposition 4.2.1 *For every graph G, we have $Z(G) = \sigma(\mathcal{L}(G))$.*

4.3   Extremal problems in general graphs and trees

The monotonicity properties of Lemma 4.2.1 immediately yield the following basic theorem (cf. Proposition 1.7.1):

Theorem 4.3.1 • *For every graph G with n vertices, we have*

$$n + 1 = \sigma(K_n) \leq \sigma(G) \leq \sigma(E_n) = 2^n,$$

where K_n and E_n are the complete and edgeless graph, respectively. Equality in the first inequality only holds if G is complete, and equality in the second inequality only holds if G is edgeless.

- *For every graph G with n vertices, we have*

$$1 = Z(E_n) \leq Z(G) \leq Z(K_n) = \sum_{0 \leq k \leq n/2} \frac{n!}{2^k k!(n-2k)!},$$

again with equality only if G is edgeless (complete, respectively).

Proof:

The inequalities follow immediately from the fact that removing edges increases $\sigma(G)$ while it decreases $Z(G)$, as stated in Lemma 4.2.1. The formulas for the complete and empty graph are relatively straightforward: we have $\sigma(E_n) = 2^n$ since all vertex subsets are independent in the edgeless graph, and $Z(E_n) = 1$ since there is only one edge subset (the empty set), thus also only one matching.

The identity $\sigma(K_n) = n+1$ is also easy: note that there are no independent sets of two or more vertices, thus only n single-vertex independent sets and the empty set, for a total of $n + 1$. The most interesting formula is that for $Z(K_n)$: let k be the number of edges in a matching. There are $\binom{n}{2k}$ possibilities to choose the $2k$ vertices involved in the matching. Thereafter, there are

$$(2k - 1)!! = 1 \cdot 3 \cdot 5 \cdot (2k - 1)$$

possible matchings: we number the vertices from 1 to $2k$. Now there are $2k - 1$ choices for the edge that covers vertex 1. Next, there are $2k - 3$ choices for the edge that covers the next-smallest vertex, and so forth. Finally, we notice that

$$\binom{n}{2k}(2k - 1)!! = \frac{n!}{2^k k!(n - 2k)!},$$

and the formula for $Z(K_n)$ follows.   □

Things get somewhat more interesting if one considers only connected graphs. The edge removal argument of the previous proof shows immediately that the maximum of the Merrifield-Simmons index and the minimum of the Hosoya index can only be attained by trees (see again Proposition 1.7.1), so it will suffice to study trees. We will also consider the dual problems for trees, i.e., minimizing the Merrifield-Simmons index and maximizing the Hosoya index. As in many other instances throughout this book, the star and the path are extremal.

In the following, Fibonacci numbers play an important role. We use the notation F_n for the Fibonacci numbers, with $F_0 = 0$, $F_1 = 1$ and $F_n = F_{n-1} + F_{n-2}$ for $n \geq 2$. Consider the path P_n with n vertices: it is easy to see that $\sigma(P_1) = 2 = F_3$ and $\sigma(P_2) = 3 = F_4$ as well as $Z(P_1) = 1 = F_2$ and $Z(P_2) = 2 = F_3$. Moreover, Corollary 4.2.1, applied to one of the ends of P_n, shows that

$$\sigma(P_n) = \sigma(P_{n-1}) + \sigma(P_{n-2}) \qquad \text{and} \qquad Z(P_n) = Z(P_{n-1}) + Z(P_{n-2}).$$

It follows immediately that

$$\sigma(P_n) = F_{n+2} \qquad \text{and} \qquad Z(P_n) = F_{n+1}. \tag{4.1}$$

The Lucas numbers L_n are closely related to the Fibonacci numbers: they satisfy the same recursion (i.e., $L_n = L_{n-1} + L_{n-2}$), but with different initial values: $L_0 = 2$ and $L_1 = 1$. It is easily verified by induction that

$$L_n = F_{n+1} + F_{n-1}$$

for all positive integers n. The Merrifield-Simmons index and Hosoya index of the cycle graph C_n are equal to Lucas numbers: to see why this is so, apply the second item of Lemma 4.2.2 to one of the vertices of the cycle to obtain

$$\sigma(C_n) = \sigma(P_{n-1}) + \sigma(P_{n-3}) = F_{n+1} + F_{n-1} = L_n.$$

Since the line graph of the cycle is again a cycle with the same number of vertices, we also have (by Proposition 4.2.1)

$$Z(C_n) = \sigma(\mathcal{L}(C_n)) = \sigma(C_n) = L_n.$$

The fact that $\sigma(P_n)$ is always a Fibonacci number is also the reason why the number of independent sets was called "Fibonacci number of a graph" by Prodinger and Tichy in their first paper on the subject in [89]. The following results on trees can already be found in their paper as well:

Theorem 4.3.2 *For every tree T with n vertices, we have*

$$\sigma(T) \leq \sigma(S_n) = 2^{n-1} + 1,$$

with equality if and only if T is a star. Moreover, we have

$$\sigma(T) \geq \sigma(P_n) = F_{n+2},$$

with equality if and only if T is a path.

Proof:

We prove both inequalities by induction on n, starting with the upper bound. For $n \leq 2$, there is nothing to prove since there is only one possibly tree (up to isomorphism). Now let T be a tree with n vertices, where $n > 2$. Pick any leaf v of T, and let w be its neighbor. By Corollary 4.2.1, we have

$$\sigma(T) = \sigma(T - v) + \sigma(T - \{v, w\}).$$

The induction hypothesis yields $\sigma(T-v) \leq 2^{n-2}+1$, with equality if and only if $T - v$ is a star. Moreover, by Theorem 4.3.1 we have $\sigma(T - \{v, w\}) \leq 2^{n-2}$, with equality if and only if $T - \{v, w\}$ is an edgeless graph. Combining the two inequalities, we get

$$\sigma(T) = \sigma(T - v) + \sigma(T - \{v, w\}) \leq 2^{n-2} + 1 + 2^{n-2} = 2^{n-1} + 1,$$

and equality can only hold if $T - v$ is a star and $T - \{v, w\}$ is edgeless. It is easy to see that this is only possible if T is a star.

Now we consider the lower bound. Again, the small cases are trivial, and we can continue by induction. Let T be a tree with n vertices, $n > 2$, and let v be one of its leaves and w its neighbor as before. The induction hypothesis gives us $\sigma(T-v) \geq F_{n+1}$, with equality if and only if $T-v$ is a path. The graph $T - \{v, w\}$ is not necessarily a tree, since it might not be connected. However, it is a forest (i.e., acyclic) and can thus be obtained by removing edges from a tree. So we can combine Lemma 4.2.1 with the induction hypothesis to obtain $\sigma(T - \{v, w\}) \geq F_n$, again with equality if and only if $T - \{v, w\}$ is a path. Putting the two inequalities together, we find that

$$\sigma(T) = \sigma(T - v) + \sigma(T - \{v, w\}) \geq F_{n+1} + F_n = F_{n+2},$$

and $T - v$ and $T - \{v, w\}$ both need to be paths for equality to hold. It is easy to see that this is the case if and only if T is a path.   □

The analogous theorem for the Hosoya index is very similar, as is its proof:

Theorem 4.3.3 *For every tree T with n vertices, we have*

$$Z(T) \leq Z(P_n) = F_{n+1},$$

with equality if and only if T is a path. Moreover, we have

$$Z(T) \geq Z(S_n) = n,$$

with equality if and only if T is a star.

Proof:

For the first statement, we can copy the proof of the previous theorem: again, there is nothing to prove for $n \leq 2$. For the induction step, we consider a tree T with n vertices, where $n > 2$. Pick a leaf v, let w be the neighbor, and note that we have $Z(T - v) \leq F_n$ by the induction hypothesis, as well as $Z(T - \{v, w\}) \leq F_{n-1}$ by the induction hypothesis, combined with Lemma 4.2.1. Now the desired inequality follows from Corollary 4.2.1, and the argument that the path is the only tree for which equality holds is identical to the previous theorem.

The proof of the lower bound is even simpler. A tree with n vertices has $n - 1$ edges and therefore at least n matchings: the empty set and $n - 1$ single-edge matchings. The only trees without further matchings containing two or more edges are stars, since any two edges have a common vertex. In fact, stars and the complete graph K_3 are easily seen to be the only graphs with this property.   □

Every connected graph contains a spanning tree as a subgraph. Thus, in view of Lemma 4.2.1, the inequalities

$$\sigma(G) \leq \sigma(S_n) \qquad \text{and} \qquad Z(G) \geq Z(S_n)$$

hold for arbitrary connected graphs G, not only for trees—again, we refer to the general statement of Proposition 1.7.1.

As in Theorem 4.3.1, we observe that the graphs that yield the maximum of the Merrifield-Simmons index are those that yield the minimum of the Hosoya index, and vice versa. This is a very typical situation that we will observe repeatedly throughout this chapter.

4.4 Graph transformations

In previous chapters, various transformations of graphs were a common element in our proofs. There are a number of standard transformations that occur very frequently in the literature on the Merrifield-Simmons index and the Hosoya index, and the aim of this section is to discuss some of them with selected applications. One common transformation is the replacement of a subgraph, in particular when it comes to trees and tree-like graphs. Since the star and the path are the extremal trees with respect to the Merrifield-Simmons index and the Hosoya index, the following result might not come as a surprise:

Lemma 4.4.1 *Suppose that a graph G can be decomposed into a connected graph H and a tree T that only share a cutvertex v. Let G_1 be the graph obtained by replacing T by a path with v as one of its endpoints, and let G_2 be the graph obtained by replacing T by a star with v at the center. We have*

$$\sigma(G_1) \leq \sigma(G) \leq \sigma(G_2),$$

and equality only holds if G is isomorphic to G_1 or G_2, respectively. Likewise,

$$Z(G_2) \leq Z(G) \leq Z(G_1),$$

and again equality only holds if G is isomorphic to G_1 or G_2, respectively.

Proof:

We apply Lemma 4.2.2 to vertex v to obtain

$$\sigma(G) = \sigma(H - v)\sigma(T - v) + \sigma(H - N[v])\sigma(T - N[v])$$
$$= \sigma(H - N[v])\sigma(T) + \big(\sigma(H - v) - \sigma(H - N[v])\big)\sigma(T - v).$$

Now note that $\sigma(H - N[v])$ and $\sigma(H - v) - \sigma(H - N[v])$ are both non-negative; they are even strictly positive unless H only consists of a single vertex.

The value of $\sigma(T)$ attains its maximum (among all possible choices of a tree T) when T is a star, and the value of $\sigma(T - v)$ attains its maximum when T is a star and v its center (so that $T - v$ is edgeless). Thus the inequality

$\sigma(G) \le \sigma(G_2)$ follows. If H has more than one vertex, the inequality is strict unless T is a star with v at its center, so that G is isomorphic to G_2. If H only has a single vertex, then T just has to be a star for equality to hold, which also means that G is isomorphic to G_2.

Likewise, the value of $\sigma(T)$ attains its minimum when T is a path, and the value of $\sigma(T - v)$ attains its minimum when T is a path with v at one of its ends (so that $T - v$ is also a path). Now the inequality $\sigma(G_1) \le \sigma(G)$ follows. Again, the inequality is strict if H has more than one vertex unless T is a path with v at one of its ends. In this case, G is isomorphic to G_1, and the same is true if H only has a single vertex, so that T only has to be a path for equality to hold.

The proof of the inequalities for the Hosoya index is analogous and left as an exercise. $\qquad\square$

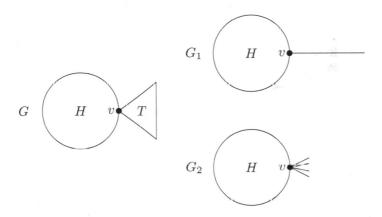

FIGURE 4.2
The transformations of Lemma 4.4.1.

For our next result, we need properties of the Fibonacci numbers. Let $\phi = \frac{1+\sqrt{5}}{2}$ be the golden ratio, and let $\overline{\phi} = -\frac{1}{\phi} = \frac{1-\sqrt{5}}{2}$. It is well known that the Fibonacci numbers satisfy the explicit formula (known as Binet's formula)

$$F_n = \frac{1}{\sqrt{5}}\left(\phi^n - \overline{\phi}^n\right).$$

The Lucas numbers L_n satisfy a very similar formula:

$$L_n = \phi^n + \overline{\phi}^n.$$

These formulas can also be used to define F_n and L_n for negative integers n: we have $F_{-n} = (-1)^{n-1}F_n$ and $L_{-n} = (-1)^n L_n$.

Lemma 4.4.2 *For every positive integer n, we have*

$$F_k F_{n-k} > F_\ell F_{n-\ell}$$

for all odd values of k and even values of ℓ with $0 \leq k, \ell \leq n/2$. Moreover, $F_k F_{n-k}$ is decreasing in k for odd $k \leq n/2$, and $F_\ell F_{n-\ell}$ is increasing in ℓ for even $\ell \leq n/2$. In other words,

$$F_1 F_{n-1} > F_3 F_{n-3} > F_5 F_{n-5} > \cdots > F_{\lfloor n/2 \rfloor} F_{\lceil n/2 \rceil} > \cdots > F_2 F_{n-2} > F_0 F_n.$$

Proof:

Direct calculation shows that

$$
\begin{aligned}
F_k F_{n-k} &= \frac{1}{5} \left(\phi^n - \overline{\phi}^k \phi^{n-k} - \overline{\phi}^{n-k} \phi^k + \overline{\phi}^n \right) \\
&= \frac{1}{5} \left(\phi^n + \overline{\phi}^n \right) - \frac{1}{5} (-1)^k \left(\phi^{n-2k} + \overline{\phi}^{n-2k} \right) \\
&= \frac{1}{5} \left(L_n + (-1)^{k+1} L_{n-2k} \right).
\end{aligned}
$$

Since L_{n-2k} is decreasing as a function of k for $k \leq n/2$ and $(-1)^{k+1} L_{n-2k}$ is positive for odd k and negative for even k, the chain of inequalities follows immediately.   □

Lemma 4.4.3 *Let G be a fixed connected graph with more than one vertex, and let v be one of its vertices. The graph $P(n, k, G, v)$ is obtained by identifying the k-th vertex of an n-vertex path with v, see Figure 4.3. Writing n as $n = 4m + i$, where $i \in \{1, 2, 3, 4\}$, we have the following chains of inequalities, where $l = \lfloor \frac{i-1}{2} \rfloor$:*

$$
\begin{aligned}
\sigma(P(n, 2, G, u)) &> \sigma(P(n, 4, G, u)) > \cdots > \sigma(P(n, 2m + 2l, G, u)) > \\
\sigma(P(n, 2m + 1, G, u)) &> \cdots > \sigma(P(n, 3, G, u)) > \sigma(P(n, 1, G, u)),
\end{aligned}
$$

and

$$
\begin{aligned}
Z(P(n, 2, G, u)) &< Z(P(n, 4, G, u)) < \cdots < Z(P(n, 2m + 2l, G, u)) < \\
Z(P(n, 2m + 1, G, u)) &< \cdots < Z(P(n, 3, G, u)) < Z(P(n, 1, G, u)).
\end{aligned}
$$

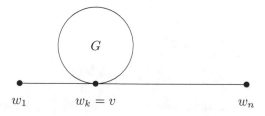

FIGURE 4.3
Illustration of the graph in Lemma 4.4.3.

Proof:

We apply the second part of Lemma 4.2.2 to the vertex v. Note that $P(n, k, G, v) - v$ decomposes into three components: $G - v$ and two paths of $k - 1$ and $n - k$ vertices, respectively. Likewise, $P(n, k, G, v) - N[v]$ decomposes into $G - N[v]$ and two paths of $k - 2$ and $n - k - 1$ vertices. Therefore,

$$\begin{aligned}
\sigma&(P(n, k, G, v)) \\
&= \sigma(P_{k-1})\sigma(P_{n-k})\sigma(G - v) + \sigma(P_{k-2})\sigma(P_{n-k-1})\sigma(G - N[v]) \\
&= \sigma(P_n)\sigma(G - N[v]) + \sigma(P_{k-1})\sigma(P_{n-k})\big(\sigma(G - v) - \sigma(G - N[v])\big) \\
&= F_{n+2}\sigma(G - N[v]) + F_{k+1}F_{n-k+2}\big(\sigma(G - v) - \sigma(G - N[v])\big).
\end{aligned}$$

Since G has more than one vertex and is connected, $G - N[v]$ is a proper subgraph of $G - v$, hence we have $\sigma(G - v) - \sigma(G - N[v]) > 0$. Therefore, the first set of inequalities follows directly from Lemma 4.4.2.

The proof for the Hosoya index follows the same lines. First we find an expression for $Z(P(n, k, G, v))$. Applying the third part of Lemma 4.2.3 to the edges of the path that are incident to v, we find that

$$\begin{aligned}
Z(P(n, k, G, v)) &= Z(P_{k-1})Z(P_{n-k})Z(G) + Z(P_{k-2})Z(P_{n-k})Z(G - v) \\
&\quad + Z(P_{k-1})Z(P_{n-k-1})Z(G - v).
\end{aligned}$$

In the special case where G is a single vertex, we obtain

$$Z(P_n) = Z(P_{k-1})Z(P_{n-k}) + Z(P_{k-2})Z(P_{n-k}) + Z(P_{k-1})Z(P_{n-k-1}),$$

so

$$\begin{aligned}
Z(P(n, k, G, v)) &= Z(P_n)Z(G - v) + Z(P_{k-1})Z(P_{n-k})\big(Z(G) - Z(G - v)\big) \\
&= F_{n+1}Z(G - v) + F_k F_{n-k+1}\big(Z(G) - Z(G - v)\big).
\end{aligned}$$

The second set of inequalities now follows in the same way as before from Lemma 4.4.2. □

As a first application of these transformations, we determine the trees with second-smallest and second largest Merrifield-Simmons index and Hosoya index, respectively.

Theorem 4.4.1 *For every positive integer $n \geq 6$, the unique tree with the second-smallest Merrifield-Simmons index among n-vertex trees is the tree resulting from attaching a path of length 2 to the third vertex of a path of $n - 2$ vertices (see Figure 4.4). The same tree is also the unique tree with the second-largest Hosoya index among n-vertex trees.*

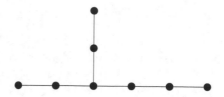

FIGURE 4.4

The tree with second-smallest Merrifield-Simmons index and second-largest Hosoya index for $n = 8$.

Proof:

Since the arguments are completely identical for the two statements, we combine the two proofs. Let $n \geq 6$ be fixed, and suppose that T is a tree with n vertices and either the property that its Merrifield-Simmons index is second-smallest among n-vertex trees (thus smallest among n-vertex trees that are not paths) or the property that its Hosoya index is second-largest among n-vertex trees (largest among n-vertex trees that are not paths).

Suppose first that there is a vertex v of degree greater than 3. If we take the tree induced by v and two of v's branches and replace it by a path, then the Merrifield-Simmons index decreases and the Hosoya index increases, both by Lemma 4.4.1. Since v still has degree at least 3 in the resulting tree, we do not obtain a path. This contradicts the choice of T.

Also, if there is a vertex v of degree 3 for which (at least) one of the branches of v is not a path, then we replace this branch by a path. The resulting tree is not a path, and the Merrifield-Simmons index decreases while the Hosoya index increases. Again, we get a contradiction.

Since every tree that is not a path contains a vertex of degree at least 3, the only remaining possibility is that there is a vertex of degree 3 for which all three branches are paths. Now assume that there are at least two branches whose length is not 2. Let the lengths of these branches be a and b. If we replace them by paths of lengths 2 and $a+b-2$, then we get another tree that is not a path. Here we may assume that a and b are not both 1: if they are, the third branch has length $n-3$, which is greater than 2, so we can take this branch instead of one of the short branches for the transformation. As before, the Merrifield-Simmons index decreases and the Hosoya index increases, this time by Lemma 4.4.3. Since this contradicts the choice of T again, we conclude that T must have two branches of length 2 (and one branch of length $n-5$), i.e., T is the tree described in the statement of the theorem. This completes our proof.   □

The dual result follows a similar line of reasoning:

Theorem 4.4.2 *For every positive integer $n \geq 4$, the unique tree with the second-smallest Merrifield-Simmons index among n-vertex trees is the tree* resulting from a star with $n - 1$ vertices by subdividing one of its edges (thus turning it into two edges), see Figure 4.5. The same tree is also the unique tree with the second-smallest Hosoya index among n-vertex trees.

FIGURE 4.5
The tree with second-largest Merrifield-Simmons index and second-smallest Hosoya index for $n = 6$.

Proof:

Again, the proofs for Merrifield-Simmons index and Hosoya index are largely identical. Let $n \geq 4$ be fixed, and suppose that T is a tree with n vertices and either the property that its Merrifield-Simmons index is second-largest among n-vertex trees (thus largest among n-vertex trees that are not stars) or the property that its Hosoya index is second-smallest among n-vertex trees (smallest among n-vertex trees that are not stars).

Suppose first that there is a non-leaf vertex v with the property that at least two of its branches are not single leaves. Pick any of these branches and replace the subtree induced by v and this branch by a star, centered at v. The resulting tree is not a star, and by Lemma 4.4.1 the Hosoya index decreases while the Merrifield-Simmons index increases. As in the proof of Theorem 4.4.1, this contradicts the choice of T.

Since we are assuming that T is not a star, there are at least two non-leaf vertices v and w. We now know that each of them can only have one non-leaf branch (which contains the other vertex). The path from v to w cannot contain any further vertices, since any such vertex would have two non-leaf branches (one containing v, the other w). Thus v and w are neighbors, and we find that T must be a double star (a tree with precisely two non-leaf vertices that are adjacent to each other). Suppose that v and w have a and b leaf neighbors, respectively. Clearly, $a + b = n - 2$. We can easily determine the Merrifield-Simmons index and the Hosoya index of T by means of Lemmas 4.2.2 and 4.2.3:

$$\sigma(T) = 2^{a+b} + 2^a + 2^b = 2^{n-2} + 2^a + 2^b$$

and

$$Z(T) = ab + a + b + 2 = ab + n.$$

Now simply note that $2^a + 2^b$, subject to the conditions $a + b = n - 2$ and $a, b \geq 1$, attains its maximum if and only if either $a = 1$ or $b = 1$, while ab

attains its minimum under the same conditions if and only if either $a = 1$ or $b = 1$. In both cases, we can conclude that T must indeed be the tree described in the statement of the theorem. □

While Lemma 4.4.1 deals with replacing tree parts of a graph, the following lemma considers the operation of moving branches, which will be useful in our study of different families of trees and tree-like graphs.

Lemma 4.4.4 *Let v and w be two distinct vertices of a graph H, and let J_1 and J_2 be two connected graphs with at least two vertices each. Let u_1 and u_2 be distinguished vertices of J_1 and J_2, respectively. Let G be the graph obtained by merging v with u_1 and w with u_2, let G_v be the graph obtained by merging v with u_1 and u_2, and let G_w be the graph obtained by merging w with u_1 and u_2. At least one of the following two inequalities holds:*

$$\sigma(G_v) > \sigma(G) \qquad or \qquad \sigma(G_w) > \sigma(G).$$

Moreover, at least one of the following two inequalities holds:

$$Z(G_v) < Z(G) \qquad or \qquad Z(G_w) < Z(G).$$

See Figure 4.6 for an illustration of this lemma.

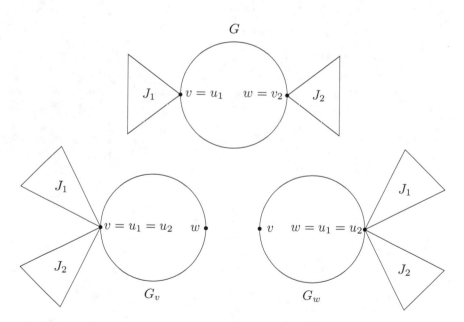

FIGURE 4.6
The graphs G, G_v and G_w in Lemma 4.4.4.

Proof:

We first determine formulas for $\sigma(G)$, $\sigma(G_v)$ and $\sigma(G_w)$. The independent sets of G can be divided into four classes, depending on whether the vertices v and w are contained or not (this amounts to applying Lemma 4.2.2 twice). For simplicity of notation, let us denote the number of independent sets of H that contain both v and w by a, b_v is the number of independent sets that contain v, but not w, b_w the number of independent sets that contain w, but not v, and finally c the number of independent sets of H that contain neither v nor w. Now we look at the graphs G, G_v and G_w:

- The number of independent sets of G that contain both v and w is

$$a\sigma(J_1 - N[u_1])\sigma(J_2 - N[u_2]),$$

  and the same applies to G_v and G_w as well.

- The number of independent sets of G that contain neither v nor w is

$$c\sigma(J_1 - u_1)\sigma(J_2 - u_2),$$

  and again this is also true for G_v and G_w.

- The number of independent sets of G that contain v, but not w, is

$$b_v\sigma(J_1 - N[u_1])\sigma(J_2 - u_2),$$

  while it is
$$b_v\sigma(J_1 - N[u_1])\sigma(J_2 - N[u_2])$$
  for G_v and
$$b_v\sigma(J_1 - u_1)\sigma(J_2 - u_2)$$
  for G_w.

- Likewise, the number of independent sets of G that contain w, but not v, is

$$b_w\sigma(J_1 - u_1)\sigma(J_2 - N[u_2]),$$

  while it is
$$b_w\sigma(J_1 - u_1)\sigma(J_2 - u_2)$$
  for G_v and
$$b_w\sigma(J_1 - N[u_1])\sigma(J_2 - N[u_2])$$
  for G_w.

We see that the difference between $\sigma(G)$, $\sigma(G_v)$ and $\sigma(G_w)$ comes from the last two cases. Combining all the formulas, we find that

$$\sigma(G_v) - \sigma(G) = \big(b_w\sigma(J_1-u_1)-b_v\sigma(J_1-N[u_1])\big)\big(\sigma(J_2-u_2)-\sigma(J_2-N[u_2])\big)$$

and

$$\sigma(G_w)-\sigma(G) = \big(b_v\sigma(J_2-u_2)-b_w\sigma(J_2-N[u_2])\big)\big(\sigma(J_1-u_1)-\sigma(J_1-N[u_1])\big).$$

Since J_1 and J_2 were assumed to be connected and to have at least two vertices each, we have $\sigma(J_1 - u_1) > \sigma(J_1 - N[u_1])$ and $\sigma(J_2 - u_2) > \sigma(J_2 - N[u_2])$. Thus at least one of the two factors $b_w\sigma(J_1-u_1)-b_v\sigma(J_1-N[u_1])$ (positive if $b_w \geq b_v$) and $b_v\sigma(J_2 - u_2) - b_w\sigma(J_2 - N[u_2])$ (positive if $b_v \geq b_w$) is positive, and the other factors $\sigma(J_1-u_1)-\sigma(J_1-N[u_1])$ and $\sigma(J_2-u_2)-\sigma(J_2-N[u_2])$ are always positive. It follows that at least one of the two inequalities $\sigma(G_v) > \sigma(G)$ and $\sigma(G_w) > \sigma(G)$ must hold (note that it is in fact possible that both hold).

Now we prove the analogous statement for the Hosoya index. In analogy to the first part of the proof, let A be the number of matchings of H that cover both v and w, B_v the number of matchings that cover v, but not w, B_w the number of matchings that cover w, but not v, and finally C the number of matchings that cover neither v nor w. Now we count matchings in G, G_v and G_w:

- The number of matchings of G that cover both v and w within H is

$$AZ(J_1 - u_1)Z(J_2 - u_2),$$

and the same applies to G_v and G_w as well.

- The number of matchings of G that cover neither v nor w within H is

$$CZ(J_1)Z(J_2),$$

but in contrast to the first half of this proof, this is not the case for G_v and G_w. Indeed, the number for these two graphs is

$$C\big(Z(J_1)Z(J_2 - u_2) + Z(J_1 - u_1)Z(J_2) - Z(J_1 - u_1)Z(J_2 - u_2)\big).$$

To see why, note that J_1 and J_2 have a vertex in common (u_1 and u_2 are merged) in both G_v and G_w. Thus a matching either does not cover u_2 in J_2 (giving the term $Z(J_1)Z(J_2 - u_2)$) or does not cover u_1 in J_1 (giving the term $Z(J_1 - u_1)Z(J_2)$), but matchings for which neither is the case are counted twice, hence we subtract $Z(J_1 - u_1)Z(J_2 - u_2)$ to make up for the overcount.

- The number of matchings of G that cover v, but not w, within H is

$$B_vZ(J_1 - u_1)Z(J_2),$$

while it is

$$B_vZ(J_1 - u_1)Z(J_2 - u_2)$$

for G_v and

$$B_v\big(Z(J_1)Z(J_2 - u_2) + Z(J_1 - u_1)Z(J_2) - Z(J_1 - u_1)Z(J_2 - u_2)\big)$$

for G_w.

- Likewise, the number of matchings of G that cover w, but not v, within H is

$$B_w Z(J_1) Z(J_2 - u_2),$$

while it is

$$B_w \big(Z(J_1) Z(J_2 - u_2) + Z(J_1 - u_1) Z(J_2) - Z(J_1 - u_1) Z(J_2 - u_2) \big)$$

for G_v and

$$B_w Z(J_1 - u_1) Z(J_2 - u_2)$$

for G_w.

Again, we consider the differences:

$$Z(G) - Z(G_v)$$
$$= \big(C(Z(J_1) - Z(J_1 - u_1)) + (B_v - B_w) Z(J_1 - u_1) \big) \big(Z(J_2) - Z(J_2 - u_2) \big)$$

and analogously

$$Z(G) - Z(G_w)$$
$$= \big(C(Z(J_2) - Z(J_2 - u_2)) + (B_w - B_v) Z(J_2 - u_2) \big) \big(Z(J_1) - Z(J_1 - u_1) \big).$$

Since J_1 and J_2 were assumed to be connected and to have at least two vertices each, we have $Z(J_1) > Z(J_1 - u_1)$ and $Z(J_2) > Z(J_2 - u_2)$. It follows that $Z(G) > Z(G_v)$ if $B_v \geq B_w$ and $Z(G) > Z(G_w)$ if $B_w \geq B_v$, so again at least one of the two desired inequalities must hold. □

The following lemma is similar in nature, and it will play the same role for problems involving minimizing σ and maximizing Z that Lemma 4.4.4 plays for maximizing σ and minimizing Z.

Lemma 4.4.5 *Let v and w be two distinct vertices of a connected graph H with more than two vertices, and consider the following three graphs, where k, h, r are positive integers:*

- G *is obtained from H by adding a path of length r between v and w, a path of length k starting at v and a path of length h starting at w (the latter two end in a leaf),*

- G_v *is obtained from H by adding a path of length r between v and w and a path of length $k + h$ starting at v,*

- G_w *is obtained from H by adding a path of length r between v and w and a path of length $k + h$ starting at w.*

See Figure 4.7 for an illustration. The following two statements hold:

- *either $\sigma(G) > \sigma(G_v)$ or $\sigma(G) > \sigma(G_w)$,*

- *either $Z(G) < Z(G_v)$ or $Z(G) < Z(G_w)$.*

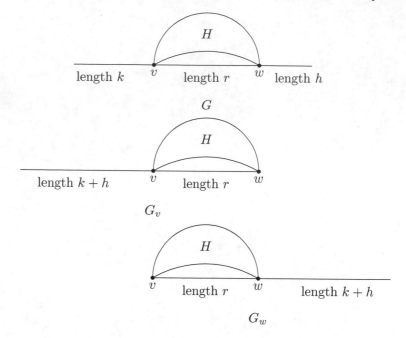

FIGURE 4.7
The graphs G, G_v and G_w in Lemma 4.4.5.

Proof:

- Let us start with the Merrifield-Simmons index. We use the same notation
  as before: the number of independent sets of H that contain both v and w
  is denoted by a, b_v is the number of independent sets that contain v, but
  not w, b_w the number of independent sets that contain w, but not v, and
  finally c the number of independent sets of H that contain neither v nor
  w. Applying Lemma 4.2.2 to v and w and the formula for the number of
  independent sets of a path from (4.1), we get

$$\sigma(G) = a\sigma(P_{k-1})\sigma(P_{h-1})\sigma(P_{r-3}) + b_v\sigma(P_{k-1})\sigma(P_h)\sigma(P_{r-2})$$
$$+ b_w\sigma(P_k)\sigma(P_{h-1})\sigma(P_{r-2}) + c\sigma(P_k)\sigma(P_h)\sigma(P_{r-1})$$

as well as

$$\sigma(G_v) = a\sigma(P_{k+h-1})\sigma(P_{r-3}) + b_v\sigma(P_{k+h-1})\sigma(P_{r-2})$$
$$+ b_w\sigma(P_{k+h})\sigma(P_{r-2}) + c\sigma(P_{k+h})\sigma(P_{r-1})$$

and

$$\sigma(G_w) = a\sigma(P_{k+h-1})\sigma(P_{r-3}) + b_v\sigma(P_{k+h})\sigma(P_{r-2})$$
$$+ b_w\sigma(P_{k+h-1})\sigma(P_{r-2}) + c\sigma(P_{k+h})\sigma(P_{r-1}).$$

We take the differences, recall that $\sigma(P_n) = F_{n+2}$ and apply the identities $F_{k+1}F_{h+1} + F_kF_h = F_{k+h+1}$ and $F_{k+2}F_{h+2} - F_kF_h = F_{k+h+2}$ that follow readily from Binet's formula (cf. Lemma 4.4.2) or combinatorially (Exercise 5). This gives us

$$\sigma(G) - \sigma(G_v) = -aF_kF_hF_{r-1} + b_vF_{k-1}F_hF_r - b_wF_{k+1}F_hF_r + cF_kF_hF_{r+1}$$

and analogously

$$\sigma(G) - \sigma(G_w) = -aF_kF_hF_{r-1} - b_vF_kF_{h+1}F_r + b_wF_kF_{h-1}F_r + cF_kF_hF_{r+1}.$$

We remark that these equations even remain correct (as one can check easily) when r is so small that expressions such as $\sigma(P_{r-3})$ are not well defined (Exercise 6). This also applies later to the proof for the Hosoya index. Using the recursion $F_{k+1} = F_k + F_{k-1}$ (and likewise $F_{h+1} = F_h + F_{h-1}$), we can rewrite these as

$$\sigma(G) - \sigma(G_v) = (cF_{r+1} - b_wF_r - aF_{r-1})F_kF_h + (b_v - b_w)F_{k-1}F_hF_r$$

and

$$\sigma(G) - \sigma(G_w) = (cF_{r+1} - b_vF_r - aF_{r-1})F_kF_h + (b_w - b_v)F_kF_{h-1}F_r.$$

Now note that $c > b_v$, $c > b_w$ and $c > a$: for every independent set that contains v, but not w, we can simply remove v to obtain an independent set that contains neither v nor w. This is an injective relation, so $c \geq b_v$. Moreover, the inequality must be strict: there are independent sets containing neither v nor w that cannot be obtained in this way (all those that contain neighbors of v), since H is connected and has at least three vertices by our assumptions. Thus $c > b_v$, and the inequalities $c > b_w$ and $c > a$ follow analogously. It follows that

$$cF_{r+1} - b_vF_r - aF_{r-1} > cF_{r+1} - cF_r - cF_{r-1} = 0,$$

which means that $\sigma(G) > \sigma(G_v)$ if $b_v \geq b_w$, and $\sigma(G) > \sigma(G_w)$ if $b_w \geq b_v$. Clearly, one of these two inequalities must hold.

- The proof of the second statement is very similar. We let A be the number of matchings of H that cover both v and w, B_v the number of matchings that cover v, but not w, B_w the number of matchings that cover w, but not v, and finally C the number of matchings that cover neither v nor w. Now we count matchings in G, G_v and G_w again, in the same way as in the proof of Lemma 4.4.4.

  - The number of matchings of G that cover both v and w within H is

  $$AZ(P_k)Z(P_h)Z(P_{r-1}) = AF_{k+1}F_{h+1}F_r,$$

  while the corresponding number for G_v and G_w is

  $$AZ(P_{k+h})Z(P_{r-1}) = AF_{k+h+1}F_r.$$

- The number of matchings of G that cover neither v nor w within H is

$$CZ(P_{k+h+r+1}) = CF_{k+h+r+2},$$

and the number remains the same for G_v and G_w.

- The number of matchings of G that cover v, but not w, within H is

$$B_v Z(P_k)Z(P_{h+r}) = B_v F_{k+1}F_{h+r+1},$$

while it is

$$B_v Z(P_{k+h})Z(P_r) = B_v F_{k+h+1}F_{r+1}$$

for G_v and

$$B_v Z(P_{k+h+r}) = B_v F_{k+h+r+1}$$

for G_w.

- Likewise, the number of matchings of G that cover w, but not v, within H is

$$B_w Z(P_h)Z(P_{k+r}) = B_w F_{h+1}F_{k+r+1},$$

while it is

$$B_w Z(P_{k+h+r}) = B_w F_{k+h+r+1}$$

for G_v and

$$B_w Z(P_{k+h})Z(P_r) = B_w F_{k+h+1}F_{r+1}$$

for G_w.

In calculating the differences, we use the identity $F_{k+1}F_{h+1}+F_k F_h = F_{k+h+1}$ that we also used in the first part, as well as the more general version $F_{k+h+1}F_{r+1} - F_{k+1}F_{h+r+1} = (-1)^r F_h F_{k-r}$ ($r = 0$ gives the aforementioned identity as a special case). These give us

$$\begin{aligned}
Z(G_v) - Z(G) &= AF_r(F_{k+h+1} - F_{k+1}F_{h+1}) + B_v(F_{k+h+1}F_{r+1} \\
&\quad - F_{k+1}F_{h+r+1}) + B_w(F_{k+h+r+1} - F_{h+1}F_{k+r+1}) \\
&= AF_r F_k F_h + (-1)^r B_v F_h F_{k-r} + B_w F_h F_{k+r} \\
&= F_h\big(AF_r F_k + (-1)^r B_v F_{k-r} + B_w F_{k+r}\big)
\end{aligned}$$

and analogously

$$Z(G_w) - Z(G) = F_k\big(AF_r F_h + (-1)^r B_w F_{h-r} + B_v F_{h+r}\big).$$

Since we are assuming that k, h, r are positive integers, we have $F_k, F_h > 0$. Moreover, as in the proof of Lemma 4.4.4 we have $B_v, B_w > 0$ by our assumptions on H. Finally, $|(-1)^r F_{k-r}| = F_{|k-r|} < F_{k+r}$ since $|k - r| \leq k + r - 2$. Thus we can conclude that

$$Z(G_v) - Z(G) \geq F_h(-B_v F_{|k-r|} + B_w F_{k+r}) > (B_w - B_v)F_h F_{k+r}$$

and by the same reasoning

$$Z(G_w) - Z(G) \geq F_k(B_v F_{h+r} - B_w F_{|h-r|}) > (B_v - B_w)F_k F_{h+r}.$$

Clearly, it follows that at least one of the two differences is positive, which is what we wanted to prove.　　□

Our next lemma also deals with a similar situation, but allows for more invasive transformations. In order to formulate it, we need two invariants associated with rooted graphs that are derived from the Merrifield-Simmons index and the Hosoya index.

Let G be a rooted graph (i.e., a graph with a distinguished vertex, called the root), and let v be its root. We set

$$\rho(G) = \frac{\sigma(G - v)}{\sigma(G)} \qquad \text{and} \qquad \tau(G) = \frac{Z(G - v)}{Z(G)}.$$

In words, $\rho(G)$ is the proportion of independent sets of G that do not contain the root v, and $\tau(G)$ is the proportion of matchings of G that do not cover the root. In the following, it will be useful to extend the definition of σ, Z, ρ and τ to the empty graph $\emptyset$ without vertices and edges by setting $\sigma(\emptyset) = 1$, $Z(\emptyset) = 1$, $\rho(\emptyset) = 1$ and $\tau(\emptyset) = 0$. These choices will be motivated later.

Lemma 4.4.6 *Let v and w be two distinct vertices of a graph H, and let $J_1, J_2, \ldots, J_{2r}$ be rooted graphs with roots $u_1, u_2, \ldots, u_{2r}$, respectively; these graphs may also be empty (in which case there is no root). For a partition of $\{1, 2, \ldots, 2r\}$ into two disjoint sets V and W such that $V \cup W = \{1, 2, \ldots, 2r\}$, construct the graph $G_{V,W}$ as follows: take the union of H and $J_1, J_2, \ldots, J_{2r}$. Then connect v by an edge to the roots u_i of all non-empty graphs J_i with $i \in V$, and connect w by an edge to the roots u_i of all non-empty graphs J_i with $i \in W$ (see Figure 4.8). The following statements hold:*

- *Among all possible partitions of $\{1, 2, \ldots, 2r\}$ into disjoint sets V and W with $|V| = |W| = r$, the maximum of $\sigma(G_{V,W})$ can only be attained if one of the following holds:*

$$\min_{i \in V} \rho(J_i) \geq \max_{i \in W} \rho(J_i)$$

  or

$$\max_{i \in V} \rho(J_i) \leq \min_{i \in W} \rho(J_i).$$

  In words: the graphs J_i with the r largest values of $\rho(J_i)$ or the graphs J_i with the r smallest values of $\rho(J_i)$ are attached to v, the rest to w.

- *Among all possible partitions of $\{1, 2, \ldots, 2r\}$ into disjoint sets V and W with $|V| = |W| = r$, the minimum of $Z(G_{V,W})$ can only be attained if one of the following holds:*

$$\min_{i \in V} \tau(J_i) \geq \max_{i \in W} \tau(J_i)$$

  or

$$\max_{i \in V} \tau(J_i) \leq \min_{i \in W} \tau(J_i).$$

  In words: the graphs J_i with the r largest values of $\tau(J_i)$ or the graphs J_i with the r smallest values of $\tau(J_i)$ are attached to v, the rest to w.

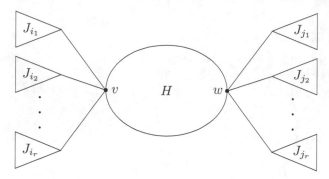

FIGURE 4.8
Illustration of the graph in Lemma 4.4.6.

Proof:

Let us first consider the proof for the Merrifield-Simmons index. We start by determining a general formula for $\sigma(G_{V,W})$. We use the same notation as in the proof of Lemma 4.4.4, where four types of independent sets are distinguished, depending on whether or not v and w are contained. Let us denote the number of independent sets of H that contain both v and w by a, b_v is the number of independent sets that contain v, but not w, b_w the number of independent sets that contain w, but not v, and finally c the number of independent sets of H that contain neither v nor w. We find that

$$\sigma(G_{V,W}) = a \prod_{i \in V} \sigma(J_i - v_i) \prod_{i \in W} \sigma(J_i - v_i) + b_v \prod_{i \in V} \sigma(J_i - v_i) \prod_{i \in W} \sigma(J_i)$$
$$+ b_w \prod_{i \in V} \sigma(J_i) \prod_{i \in W} \sigma(J_i - v_i) + c \prod_{i \in V} \sigma(J_i) \prod_{i \in W} \sigma(J_i).$$

The first term and the last term are independent of our choice of V and W: they are always equal to

$$a \prod_{i=1}^{2r} \sigma(J_i - v_i) + c \prod_{i=1}^{2r} \sigma(J_i).$$

So in order to maximize the value of $\sigma(G_{V,W})$, we have to maximize

$$b_v \prod_{i \in V} \sigma(J_i - v_i) \prod_{i \in W} \sigma(J_i) + b_w \prod_{i \in V} \sigma(J_i) \prod_{i \in W} \sigma(J_i - v_i)$$

$$= \prod_{i=1}^{2r} \sigma(J_i) \left(b_v \prod_{i \in V} \rho(J_i) + b_w \prod_{i \in W} \rho(J_i) \right).$$

We remark that this is even correct if some of the J_i are empty, since factors of $\sigma(\emptyset) = 1$ and $\rho(\emptyset) = 1$ do not affect the respective products. The products

$S = \prod_{i=1}^{2r} \sigma(J_i)$ and $R = \prod_{i=1}^{2r} \rho(J_i)$ are both independent of the choice of V and W. Setting $x = \prod_{i \in V} \rho(J_i)$, we have to maximize

$$S\left(b_v x + b_w \frac{R}{x}\right).$$

This is a convex function of x (since its second derivative is $\frac{2b_w RS}{x^3} > 0$), so its maximum is attained when x either reaches its maximum or its minimum value. The former is clearly the case if and only if

$$\min_{i \in V} \rho(J_i) \geq \max_{i \in W} \rho(J_i),$$

while the latter happens if and only if

$$\max_{i \in V} \rho(J_i) \leq \min_{i \in W} \rho(J_i).$$

This settles the lemma for the Merrifield-Simmons index. For the Hosoya index, we proceed similarly. Let A be the total number of matchings of H, B_v the number of matchings that do not cover v (but may or may not cover w), B_w the number of matchings that do not cover w (but may or may not cover v), and finally C the number of matchings of H that cover neither v nor w. We have

$$Z(G_{V,W})$$
$$= A \prod_{i \in V} Z(J_i) \prod_{i \subset W} Z(J_i) + B_v \prod_{i \in W} Z(J_i) \left(\sum_{i \in V} Z(J_i - v_i) \prod_{j \in V \setminus \{i\}} Z(J_j)\right)$$
$$+ B_w \prod_{i \in V} Z(J_i) \left(\sum_{i \in W} Z(J_i - v_i) \prod_{j \in W \setminus \{i\}} Z(J_j)\right)$$
$$+ C \left(\sum_{i \in V} Z(J_i - v_i) \prod_{j \in V \setminus \{i\}} Z(J_j)\right) \left(\sum_{i \in W} Z(J_i - v_i) \prod_{j \in W \setminus \{i\}} Z(J_j)\right).$$

Again, the first term does not depend on the choice of V and W, so it remains to minimize

$$B_v \prod_{i \in W} Z(J_i) \left(\sum_{i \in V} Z(J_i - v_i) \prod_{j \in V \setminus \{i\}} Z(J_j)\right)$$
$$+ B_w \prod_{i \in V} Z(J_i) \left(\sum_{i \in W} Z(J_i - v_i) \prod_{j \in W \setminus \{i\}} Z(J_j)\right)$$
$$+ C \left(\sum_{i \in V} Z(J_i - v_i) \prod_{j \in V \setminus \{i\}} Z(J_j)\right) \left(\sum_{i \in W} Z(J_i - v_i) \prod_{j \in W \setminus \{i\}} Z(J_j)\right)$$
$$= \prod_{i=1}^{2r} Z(J_i) \left(B_v \sum_{i \in V} \tau(J_i) + B_w \sum_{i \in W} \tau(J_i) + \left(\sum_{i \in V} \tau(J_i)\right)\left(\sum_{i \in W} \tau(J_i)\right)\right).$$

Again, this remains correct if some of the J_i are empty, since factors of $Z(\emptyset) = 1$ and summands of $\tau(\emptyset) = 0$ do not affect the expression. The product $P = \prod_{i=1}^{2r} Z(J_i)$ is independent of the choice of V and W, as is the sum $T = \sum_{i=1}^{2r} \tau(J_i)$. Setting $x = \sum_{i \in V} \tau(J_i)$, we have to minimize

$$P(B_v x + B_w(T - x) + Cx(T - x)).$$

This is a concave function of x (since its second derivative is $-2PC < 0$), so it attains its minimum either when x reaches its maximum or when x reaches its minimum. The assertion of the lemma follows as before in the case of the Merrifield-Simmons index. $\qquad\square$

4.5   Trees with fixed parameters

In this section, we will be concerned with trees that satisfy some additional conditions, e.g., on the diameter or the number of leaves. The proofs of the results in this section also illustrate the use of the transformations discussed in the previous section.

Theorem 4.5.1 ([69, 121]) *For every tree T with n vertices and diameter D, we have*

$$\sigma(T) \le 2^{n-D} F_{D+1} + F_D$$

and

$$Z(T) \le (n - D)F_D + F_{D+1}.$$

In both inequalities, equality holds if and only if T is the tree obtained by attaching a path of length $D-1$ to the center of a star with $n - D + 1$ vertices.

Proof:

Both statements can be proved by the same approach. Let T be a tree with n vertices and diameter D for which either the Merrifield-Simmons index attains its maximum, or the Hosoya index attains its minimum. Next consider a diametrical path $v_0, v_1, \ldots, v_D$, i.e., a path whose length is the diameter. Clearly, v_0 and v_D have to be leaves. When the edges of this path are removed, connected components $T_1, T_2, \ldots, T_{D-1}$ containing the vertices $v_1, v_2, \ldots, v_{D-1}$ remain. If any of the T_i is not a star rooted at v_i, then by Lemma 4.4.1 we can replace it by such a star in T, increasing the Merrifield-Simmons index and decreasing the Hosoya index. Note that this does not affect the diameter: the diametrical path remains, and no longer paths can be created by the replacement. Since this yields a contradiction to the choice of T, we can assume that each T_i is a star rooted at v_i. Next observe that at most one of the T_i can be non-trivial (contain more vertices than just v_i): if T_i and T_j are both non-trivial, then by Lemma 4.4.4 we can transfer leaves from v_i to v_j or from v_j to v_i, increasing the Merrifield-Simmons index and decreasing the Hosoya index. Again, we obtain a contradiction.

It only remains to determine the right choice of i for which T_i is a non-trivial star (with $n - D - 1$ vertices). However, this is again easy: by Lemma 4.4.3, the maximum of the Merrifield-Simmons index and the minimum of the Hosoya index are both attained when $i = 1$ or $i = D - 1$. The two choices are equivalent since they both yield the tree described in the statement of the theorem. The formulas for the Merrifield-Simmons index and the Hosoya index are obtained by means of Lemma 4.2.2 and Lemma 4.2.3. □

The tree that occurred in Theorem 4.5.1 (a "comet", see Section 1.3) also maximizes the Merrifield-Simmons index and minimizes the Hosoya index when the number of leaves is fixed. This is stated in our next theorem.

Theorem 4.5.2 ([86, 124]) *For every tree T with n vertices and ℓ leaves, we have*
$$\sigma(T) \leq 2^{\ell-1} F_{n-\ell+2} + F_{n-\ell+1}$$
and
$$Z(T) \geq (\ell - 1) F_{n-\ell+1} + F_{n-\ell+2}.$$
In both inequalities, equality holds if and only if T is the tree obtained by attaching a path of length $n - \ell$ to the center of a star with ℓ vertices.

Proof:

This proof is similar to the previous one. Again, consider a tree T that is extremal (this time among trees with n vertices and ℓ leaves). Suppose there are two vertices v and w in this tree that are either adjacent to two or more leaves, or to one leaf and more than one other vertex. By Lemma 4.4.4, we can either move a leaf from v to w, or from w to v to increase the Merrifield-Simmons index and decrease the Hosoya index. This does not change the number of leaves (since v and w do not become leaves through this transformation by our assumptions, the leaf set stays the same), so we obtain a contradiction to the choice of T. This means that there is at most one vertex that is adjacent to more than one leaf or to one leaf and more than one other vertex.

Now we use induction on the difference $n - \ell$ to complete the proof. If $n - \ell = 1$, then there is only one possible tree, namely the star. Thus the statement holds trivially in this case, and we can proceed with the induction step. If $n - \ell > 1$, then it is impossible that all leaves have the same neighbor. So an extremal tree T must have at least two vertices that have one or more leaf neighbors. By the observation above, we can further assume that at least one of them has only two neighbors. In the case that $n - \ell = 2$, this already characterizes the tree uniquely (as a comet), so we assume that $n - \ell > 2$. Let us denote this vertex by v, and let u be its leaf neighbor. Note that $T - u$ has $n - 1$ vertices and ℓ leaves, while $T - \{u, v\}$ has $n - 2$ vertices and $\ell - 1$ or

ℓ leaves (depending on whether v's other neighbor becomes a leaf or not). In either case, we can apply the induction hypothesis: we have

$$\sigma(T - u) \leq 2^{\ell-1} F_{n-\ell+1} + F_{n-\ell},$$

with equality if and only if $T - u$ is the comet with $n - 1$ vertices and ℓ leaves. Moreover, we obtain

$$\sigma(T - \{u, v\}) \leq 2^{\ell-1} F_{n-\ell} + F_{n-\ell-1}$$

if $T - \{u, v\}$ has ℓ leaves, and

$$\sigma(T - \{u, v\}) \leq 2^{\ell-2} F_{n-\ell+1} + F_{n-\ell}$$

if it has $\ell - 1$ leaves. Since

$$\left(2^{\ell-1} F_{n-\ell} + F_{n-\ell-1}\right) - \left(2^{\ell-2} F_{n-\ell+1} + F_{n-\ell}\right) = (2^{\ell-2} - 1) F_{n-\ell-2} \geq 0$$

with equality only for $\ell = 2$ (in which case $T - \{u, v\}$ cannot actually have $\ell - 1$ leaves), we can conclude that

$$\sigma(T - \{u, v\}) \leq 2^{\ell-1} F_{n-\ell} + F_{n-\ell-1},$$

with equality if and only if $T - \{u, v\}$ is the comet with $n - 2$ vertices and ℓ leaves. Now we apply Corollary 4.2.1 to combine the two:

$$\begin{aligned}
\sigma(T) &= \sigma(T - u) + \sigma(T - \{u, v\}) \\
&\leq \left(2^{\ell-1} F_{n-\ell+1} + F_{n-\ell}\right) + \left(2^{\ell-1} F_{n-\ell} + F_{n-\ell-1}\right) \\
&= 2^{\ell-1} F_{n-\ell+2} + F_{n-\ell+1},
\end{aligned}$$

with equality if and only if $T - u$ and $T - \{u, v\}$ are both comets with ℓ leaves. It is easy to see that this is only possible if T itself is a comet with ℓ leaves. The calculations for the Hosoya index are left as an exercise. □

Next, we consider trees where, in addition to the number of vertices, the maximum degree is fixed. For these trees, minimizing the Merrifield-Simmons index and maximizing the Hosoya index turns out to be easier.

Theorem 4.5.3 ([104]) *Let T be a tree with n vertices and maximum degree Δ. If $\Delta \geq \frac{n-1}{2}$, then the following two inequalities hold:*

$$\sigma(T) \geq 3^{n-\Delta-1} 2^{2\Delta-n+1} + 2^{n-\Delta-1}$$

and

$$Z(T) \leq 2^{n-\Delta-2}(3\Delta - n + 3).$$

Equality holds in both inequalities if and only if T is an extended star obtained by attaching $n - \Delta - 1$ paths of length 2 and $2\Delta - n + 1$ pendant edges to a common center.

If $\Delta < \frac{n-1}{2}$, then the inequalities

$$\sigma(T) \geq 3^{\Delta-1} F_{n-2\Delta+3} + 2^{\Delta-1} F_{n-2\Delta+2}$$

and

$$Z(T) \leq 2^{\Delta-2} \left((\Delta+1) F_{n-2\Delta+2} + 2 F_{n-2\Delta+1} \right)$$

hold, with equality in both inequalities if and only if T is an extended star obtained by attaching $\Delta - 1$ paths of length 2 and a single path of length $n - 2\Delta + 1$ to a common center.

Proof:

Let T be a tree with n vertices and maximum degree Δ for which $\sigma(T)$ attains its minimum or $Z(T)$ attains its maximum, let v be a vertex whose degree is Δ, and let $T_1, T_2, \ldots, T_\Delta$ be the branches of v (the connected components that result when v is removed). By Lemma 4.4.1, we know that each of the branches T_i has to be a path, for otherwise we could decrease the Merrifield-Simmons index and increase the Hosoya index by replacing it by a path, contradicting the choice of T. It remains to determine the lengths of the branches. If there are more than two branches with a length other than 1 or 2 (let us denote these lengths by k and ℓ, where $k, \ell > 2$), then we reach another contradiction, this time with Lemma 4.4.3: if we replace the two paths by a path of length 2 and another path of length $k + \ell - 2$, then the Merrifield-Simmons index decreases while the Hosoya index increases by Lemma 4.4.3 (applied to the graph G consisting of v and all other branches). Thus this is also impossible, which means that there is at most one branch whose length is greater than 2. The same argument also shows that there cannot be a branch of length 1 and another branch of length greater than 2.

This leaves us with two possibilities.

- All branches have length 1 or 2: in this case, we let r be the number of branches of length 1 and s the number of branches of length 2. The following equations must be satisfied:

$$r + s = \Delta \qquad \text{and} \qquad r + 2s = n - 1,$$

since the degree of v was assumed to be Δ, and since the total number of edges is $n - 1$. The solution to this system of equations is given by $r = 2\Delta - n + 1$ and $s = n - 1 - \Delta$. However, since r must be non-negative, this only makes sense if $\Delta \geq \frac{n-1}{2}$.

- There is a branch of length greater than 2: in this case, there is exactly one such branch (whose length we denote by r), while all other branches have length 2. The number of these branches will be denoted by s. As in the first case, we get a system of equations, namely

$$1 + s = \Delta \qquad \text{and} \qquad r + 2s = n - 1.$$

This gives us $r = n - 2\Delta + 1$ and $s = \Delta - 1$, which can only apply if $\Delta < \frac{n-1}{2}$ (since $r > 2$ by assumption).

In either case, we find that T must be exactly the tree described in the statement of the theorem. The formulas in terms of n and Δ are easily obtained by means of Lemmas 4.2.2 and 4.2.3.   □

The analogous problem of maximizing the Merrifield-Simmons index and minimizing the Hosoya index of trees with a given number of vertices and given maximum degree leads to more complicated structures. The solution can be obtained as a special case of Theorem 4.5.4 later in this section, which deals with trees whose degree sequence is given. We will briefly discuss this theorem and its consequences in the remainder of this section. The following lemma, which is a rather direct consequence of Lemma 4.4.6, is key to this result.

Lemma 4.5.1 *Suppose that T is a tree that maximizes the Merrifield-Simmons index among all trees with the same degree sequence. For every possible way of decomposing T as in Figure 4.9, one of the following two statements holds:*

- *$k \leq \ell$ and*

$$\min_i \rho(A_i) \geq \max_j \rho(B_j),$$

  or

- *$k \geq \ell$ and*

$$\max_i \rho(A_i) \leq \min_j \rho(B_j).$$

Likewise, suppose that T is a tree that minimizes the Hosoya index among all trees with the same degree sequence. For every possible way of decomposing T as in Figure 4.9, one of the following two statements holds:

- *$k \geq \ell$ and*

$$\min_i \tau(A_i) \geq \max_j \tau(B_j),$$

  or

- *$k \leq \ell$ and*

$$\max_i \tau(A_i) \leq \min_j \tau(B_j).$$

Here, ρ and τ are defined as in Lemma 4.4.6.

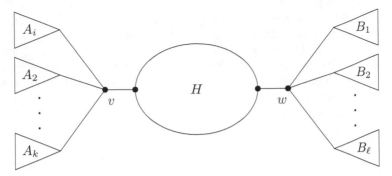

FIGURE 4.9
The decomposition of Lemma 4.5.1.

Proof:

We consider the proof for the Merrifield-Simmons index. Observe that we are exactly in the situation of Lemma 4.4.6 if we set $r = \max\{k, \ell\}$ and let $J_1, J_2, \ldots, J_{2r}$ be the branches $A_1, A_2, \ldots, A_k$ and $B_1, B_2, \ldots, B_\ell$, plus $|k - \ell|$ copies of the empty graph. Observe that

$$1 = \rho(\emptyset) > \rho(G) = \frac{\sigma(G - v)}{\sigma(G)} \tag{4.2}$$

holds for every non-empty graph G. We know from Lemma 4.4.6 that the maximum of the Merrifield-Simmons index among all possible ways of distributing the branches is attained when the r branches with the greatest values of ρ are attached to one of the two vertices v, w, and the other r branches are attached to the other. In view of (4.2), the $|k-\ell|$ empty branches have the greatest value of ρ. Therefore, we know that they belong to one part of an optimal partition, together with the $\min\{k, \ell\}$ remaining branches whose ρ-value is greatest. The other $\max\{k, \ell\}$ branches belong to the other part of the partition. Since we are assuming that T is a tree that maximizes the Merrifield-Simmons index, the branches must actually form such an optimal partition.

The artificial empty branches are necessarily attached to the vertex whose degree is smaller, together with the large ρ-values. Thus we either have $k \leq \ell$ and

$$\min_i \rho(A_i) \geq \max_j \rho(B_j),$$

or $k \geq \ell$ and

$$\max_i \rho(A_i) \leq \min_j \rho(B_j),$$

which is exactly what we wanted to prove. The proof for the Hosoya index is analogous, the only difference being that $\tau(\emptyset) = 0$ is clearly the least possible value of τ, so the artificial empty branches have to be attached to the same vertex as those $\min\{k, \ell\}$ other branches with the smallest τ-values. $\square$

A tree that satisfies the condition given in Lemma 4.5.1 either for the Merrifield-Simmons index or for the Hosoya index is called *exchange-extremal*. Andriantiana [3] showed that exchange-extremality is actually sufficient to characterize a tree completely. The result he obtained parallels Theorem 2.1.2, but with another special type of tree taking the place of the greedy tree. Since its proof is rather long and technical, it is skipped here, but let us describe his construction.

Definition 4.5.1 *Let* $(d_1, d_2, \ldots, d_k, 1, 1, \ldots, 1)$ *be a degree sequence of a tree, where* $d_k \geq 2$, *in non-increasing order. We define a tree* $\mathcal{M}(d_1, d_2, \ldots, d_k, 1, 1, \ldots, 1)$ *associated with the degree sequence by the following recursive procedure: if* $d_k \geq k - 1$, *then* $\mathcal{M}(d_1, d_2, \ldots, d_k, 1, 1, \ldots, 1)$ *is obtained from the stars* $S_{d_1}, S_{d_2}, \ldots, S_{d_{k-1}}$ *by connecting their centers to a common vertex labeled* v_1 *and attaching* $d_k - k + 1$ *leaves to* v_1 *(so that its degree becomes* d_k). *Moreover, labels* $v_2, \ldots, v_k$ *are assigned to the non-leaf neighbors of* v_1 *in increasing order of degree (i.e., the vertex labeled* v_i *has degree* d_{k+1-i}).

If $d_k < k - 1$, *then we define* $\mathcal{M}(d_1, d_2, \ldots, d_k, 1, 1, \ldots, 1)$ *as follows: let* ℓ *be the greatest integer such that* v_ℓ *occurs as a label in the tree* $\mathcal{M}(d_{d_k}, \ldots, d_{k-1}, 1, 1, \ldots, 1)$, *and let* s *be the least integer such that there is a vertex* v_s *adjacent to a leaf in that tree. Now we obtain* $\mathcal{M}(d_1, d_2, \ldots, d_k, 1, 1, \ldots, 1)$ *from* $\mathcal{M}(d_{d_k}, \ldots, d_{k-1}, 1, 1, \ldots, 1)$ *by connecting one of the leaf neighbors of* v_s *by an edge to the centers of* $d_k - 1$ *disjoint stars* $S_{d_1}, S_{d_2}, \ldots, S_{d_{d_k-1}}$. *The centers of these stars receive the labels* $v_{\ell+1}, \ldots, v_{\ell+d_k-1}$, *in increasing order of degree.*

Figure 4.10 illustrates the construction of Definition 4.5.1. The construction is very similar to that of alternating trees (see Definition 3.2.1), but yields a unique tree for every degree sequence. Note that large and small degrees alternate in the tree. In the following, we will write $\mathcal{M}(\pi)$ for the tree defined in Definition 4.5.1 associated with a degree sequence $\pi = (d_1, d_2, \ldots, d_k, 1, 1, \ldots, 1)$, and call it an $\mathcal{M}$-*tree* for short.

Theorem 4.5.4 *Every exchange-extremal tree as described in Lemma 4.5.1 is an* $\mathcal{M}$-*tree. In particular, for every possible degree sequence* π *(sequence that can be the degree sequence of a tree, cf. Section 1.4) there is a unique tree with that degree sequence that maximizes the Merrifield-Simmons index, namely the* $\mathcal{M}$-*tree* $\mathcal{M}(\pi)$ *described in Definition 4.5.1. Likewise, for every possible degree sequence* π *there is a unique tree with that degree sequence that minimizes the Hosoya index, namely the* $\mathcal{M}$-*tree* $\mathcal{M}(\pi)$.

There is also a majorization result, paralleling Theorem 2.6.5, that can be used to derive a number of corollaries from Theorem 4.5.4.

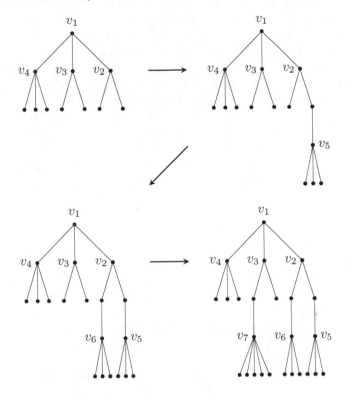

FIGURE 4.10
Construction of the tree $\mathcal{M}(6, 4, 4, 4, 3, 3, 3, 2, 2, 2, 1, 1, \ldots, 1)$.

Theorem 4.5.5 *Let π and π' be two degree sequences of trees of the same length such that π' majorizes π. If $\mathcal{M}(\pi)$ and $\mathcal{M}(\pi')$ are the $\mathcal{M}$-trees associated with π and π', then we have*

$$\sigma(\mathcal{M}(\pi')) > \sigma(\mathcal{M}(\pi))$$

and

$$Z(\mathcal{M}(\pi')) < Z(\mathcal{M}(\pi)).$$

Proof:

This result is also based on Lemma 4.4.6. We focus on the statement for the Merrifield-Simmons index, the proof for the Hosoya index is analogous. In view of Lemma 1.4.1, it suffices to prove the statement in the case where π and π' only differ in two positions, which are equal to k, ℓ (with $k \geq \ell \geq 2$) in π and $k + 1, \ell - 1$ in π'. In the $\mathcal{M}$-tree $\mathcal{M}(\pi)$, we find two vertices v and w whose degrees are k and ℓ, respectively. Now we decompose the tree $\mathcal{M}(\pi)$ into the $k - 1$ branches of v consisting of those vertices for which the unique path to w passes through v, the $\ell - 1$ branches of w consisting of those vertices for which the unique path to v passes through w, and the remaining tree H. This gives us a decomposition as in Lemma 4.4.6, where we set $r = k$ and add one artificial empty branch to v and $k - \ell + 1$ empty branches to w.

Lemma 4.4.6, combined with the observation that the empty branches always have the greatest ρ-values (see (4.2)), shows that we can obtain a new tree T with greater Merrifield-Simmons index by permuting the branches, and the result of this permutation will be a tree where all empty branches are attached to the same vertex (either v or w). Thus v and w will have degrees $k + 1$ and $\ell - 1$ (in some order), while all other degrees are the same as in $\mathcal{M}(\pi)$. Thus the new tree T has degree sequence π' (it may or may not be the $\mathcal{M}$-tree associated with π'), and we find that

$$\sigma(\mathcal{M}(\pi')) \geq \sigma(T) > \sigma(\mathcal{M}(\pi)),$$

which completes the proof.   □

It is easy to see that among all degree sequences of trees whose length is n and that contain ℓ ones (corresponding to leaves), the sequence $(\ell, 2, 2, \ldots, 2, 1, 1, \ldots, 1)$ majorizes every other sequence (cf. the analogous discussion in Section 2.6.3). The $\mathcal{M}$-tree with this degree sequence is the comet. So, Theorem 4.5.2 actually follows as a corollary of Theorem 4.5.4 and Theorem 4.5.5. Likewise, Theorems 4.5.4 and 4.5.5 also immediately yield the tree with n vertices and maximum degree Δ that maximizes the Merrifield-Simmons index and the Hosoya index, thus providing the dual to Theorem 4.5.3: the degree sequence $\pi_{n,\Delta} = (\Delta, \Delta, \ldots, \Delta, k, 1, 1, \ldots, 1)$ majorizes all other possible degree sequences whose maximum is Δ, so the extremal tree is $\mathcal{M}(\pi_{n,\Delta})$. Here, the multiplicity of Δ is $\lfloor \frac{n-2}{\Delta-1} \rfloor$, and $k \in \{1, 2, \ldots, \Delta - 1\}$ is chosen to satisfy $k \equiv n - 1 \mod (\Delta - 1)$. We leave it as an exercise to derive Theorem 4.5.1 from Theorems 4.5.4 and 4.5.5 in a similar way.

4.6   Tree-like graphs

Many of the techniques developed in the previous sections also apply to graphs that are not trees, but are similar to trees. The first step in this regard is usually to consider unicyclic graphs, i.e., graphs with a single cycle (cyclomatic number 1, cf. Section 1.4). Merrifield-Simmons index and Hosoya index of unicyclic graphs have been studied in a number of papers [23, 83, 84, 87, 109, 114, 123]. For every connected unicyclic graph, the number of edges is equal to the number of vertices. The transformation ideas of Section 4.4 will turn out to be particularly valuable again. We first consider unicyclic graphs where the unique cycle has a given length ℓ:

Theorem 4.6.1 • *Among all connected unicyclic graphs with n vertices for which the cycle has length ℓ, the unique graph that attains the maximum of the Merrifield-Simmons index is obtained by picking a vertex of the cycle of length ℓ and attaching $n - \ell$ pendant vertices to it. The same graph is also the unique unicyclic graph with n vertices and cycle length ℓ that minimizes the Hosoya index.*

• *Among all connected unicyclic graphs with n vertices for which the cycle has length ℓ, the unique graph that attains the minimum of the Merrifield-Simmons index is obtained by picking a vertex of the cycle of length ℓ and attaching a path of length $n - \ell$ (at one of its ends) to it. The same graph is also the unique unicyclic graph with n vertices and cycle length ℓ that maximizes the Hosoya index.*

Proof:

• We can combine the proofs of both assertions in the first part in one argument. Let G be a unicyclic graph with cycle length ℓ for which either the Merrifield-Simmons index reaches its maximum or the Hosoya index reaches its minimum. For a vertex v on the cycle, consider the subgraph H_v induced by all vertices that lie in the same component as v when the edges of the cycle are removed. We know from Lemma 4.4.1 that the Merrifield-Simmons index increases and the Hosoya index decreases if H_v is replaced by a star with v at its center. Thus by our choice of G, we can assume that H_v is a star, and this applies to all vertices on the cycle. Moreover, if there are two non-trivial stars (i.e., more than one vertex) H_v and H_w for distinct vertices v, w on the cycle, then by Lemma 4.4.4, we can either move H_v to w or H_w to v to increase the Merrifield-Simmons index/decrease the Hosoya index. Since this also contradicts the choice of G, there is only one non-trivial star, which means that G is indeed the graph described in the statement of the theorem.

• For the second part, the argument is very similar, but we are now using Lemma 4.4.5 instead of Lemma 4.4.4. We let G be a unicyclic graph with cycle length ℓ for which either the Merrifield-Simmons index reaches its minimum or the Hosoya index reaches its maximum. For every vertex v, we can assume that the subgraph H_v (defined in the same way as in the first part) is a path with v at one of its ends. Now assume that there are two such subgraphs that are non-trivial paths (more than one vertex), and let H_v and H_w be two such paths chosen in such a way that the distance between v and w is the smallest possible. Then all vertices on a shortest path between v and w have degree 2. Let r be the length of this path, and let k and h be the lengths of H_v and H_w, respectively. Then we find ourselves in the situation of Lemma 4.4.5. But as this lemma shows, there must now be another unicyclic graph with the same cycle length, but smaller Merrifield-Simmons index and larger Hosoya index. This contradicts our choice of G.

Thus there is only (at most) one vertex v on the cycle for which H_v consists of more than one vertex. This means that G does indeed have the form described in the statement of the theorem.

□

In order to determine the maximum and minimum of Merrifield-Simmons index and Hosoya index among all connected unicyclic graphs, all that remains now is to compare different values of ℓ. This is done in the following theorem:

Theorem 4.6.2 *Let X_n be the graph obtained from the star S_n by adding an edge between two leaves, or equivalently by attaching $n - 3$ pendant vertices to one of the vertices of the triangle graph C_3. For every connected unicyclic graph G with $n \geq 3$, we have*

$$\sigma(G) \leq \sigma(X_n) = 3 \cdot 2^{n-3} + 1,$$

with equality only if G is isomorphic to X_n (or to the cycle C_4 if $n = 4$), and

$$Z(G) \geq Z(X_n) = 2n - 2,$$

again with equality only if G is isomorphic to X_n.

Proof:

Let ℓ be the length of the unique cycle. By Theorem 4.6.1, we can assume that G is the graph consisting of a cycle of length ℓ and $n - \ell$ pendant vertices attached to one of its vertices. It is not difficult to obtain $\sigma(G)$ and $Z(G)$ in this case by means of Lemmas 4.2.2 and 4.2.3:

$$\sigma(G) = 2^{n-\ell} F_{\ell+1} + F_{\ell-1} \quad \text{and} \quad Z(G) = (n - \ell + 1)F_\ell + 2F_{\ell-1}.$$

We notice that the first expression is decreasing in ℓ:

$$\left(2^{n-\ell} F_{\ell+1} + F_{\ell-1} \right) - \left(2^{n-\ell-1} F_{\ell+2} + F_\ell \right)$$
$$= 2^{n-\ell-1}(2F_{\ell+1} - F_{\ell+2}) + (F_{\ell-1} - F_\ell)$$
$$= 2^{n-\ell-1} F_{\ell-1} - F_{\ell-2} \geq F_{\ell-1} - F_{\ell-2}$$
$$= F_{\ell-3} \geq 0$$

for $3 \leq \ell \leq n - 1$, with equality only if $\ell = 3$ and $n = 4$. Thus the maximum is obtained for $\ell = 3$, and it follows that

$$\sigma(G) \leq \sigma(X_n) = 3 \cdot 2^{n-3} + 1,$$

as claimed.

For the Hosoya index, the reasoning is similar. The expression $(n - \ell + 1)F_\ell + 2F_{\ell-1}$ is increasing in ℓ:

$$
\begin{aligned}
((n - \ell)F_{\ell+1} + 2F_\ell) &- ((n - \ell + 1)F_\ell + 2F_{\ell-1}) \\
&= (n - \ell)(F_{\ell+1} - F_\ell) + (F_\ell - 2F_{\ell-1}) \\
&= (n - \ell)F_{\ell-1} - F_{\ell-3} \geq F_{\ell-1} - F_{\ell-3} \\
&= F_{\ell-2} > 0
\end{aligned}
$$

for $3 \leq \ell \leq n - 1$. Thus the minimum is obtained for $\ell = 3$ again, and we end up with

$$Z(G) \geq Z(X_n) = 2n - 2,$$

completing the proof. $\qquad\square$

The dual statement, which is given in the following theorem, can be proven along the same lines, but we will use a more direct approach paralleling the proofs of Theorem 4.3.2 and Theorem 4.3.3.

Theorem 4.6.3 *Let Y_n be the graph obtained from the triangle graph C_3 by attaching a path of length $n - 3$ to one of its vertices. For every connected unicyclic graph G with $n \geq 3$, we have*

$$\sigma(G) \geq \sigma(C_n) = \sigma(Y_n) = L_n,$$

with equality if and only if G is isomorphic to Y_n or to the cycle C_n, and

$$Z(G) \leq Z(C_n) = L_n,$$

with equality if and only if G is isomorphic to C_n.

Proof:

We can prove both statements by induction. Both inequalities are trivial for $n = 3$ and $n = 4$: there is only one unicyclic graph with three vertices (namely the cycle C_3), and there are only two unicyclic graphs with four vertices, the cycle C_4 and the graph Y_4. The values $\sigma(C_3) = Z(C_3) = 4 = L_3$, $\sigma(C_4) = \sigma(Y_4) = Z(C_4) = 7 = L_4$ and $Z(Y_4) = 6$ are easily determined.

For the induction step, we can distinguish two cases: if G is a cycle, then there is nothing to prove. Otherwise, G has at least one pendant vertex. Pick one such vertex v, and let w be its unique neighbor. We can apply the recursions of Lemma 4.2.2 and Lemma 4.2.3 to obtain

$$\sigma(G) = \sigma(G-v) + \sigma(G-\{v,w\}) \qquad \text{and} \qquad Z(G) = Z(G-v) + Z(G-\{v,w\}).$$

We can apply the induction hypothesis to $G - v$, which is still unicyclic and connected:

$$\sigma(G - v) \geq L_{n-1} \qquad \text{and} \qquad Z(G - v) \leq L_{n-1}.$$

The graph $G - \{v, w\}$ might not be unicyclic, and it might not even be connected. However, it can clearly not contain more than one cycle, so it is possible to add edges (potentially none) to $G - \{v, w\}$ to turn it into a connected unicyclic graph G' with equally many vertices. This procedure will always increase the Hosoya index and decrease the Merrifield-Simmons index (by Lemma 4.2.1), so we also have

$$\sigma(G - \{v, w\}) \geq \sigma(G') \geq L_{n-2} \quad \text{and} \quad Z(G - \{v, w\}) \leq Z(G') \leq L_{n-2}.$$

In conclusion, we have

$$\sigma(G) \geq L_{n-1} + L_{n-2} = L_n$$

and

$$Z(G) \leq L_{n-1} + L_{n-2} = L_n,$$

as required. For the Hosoya index, equality can only hold if both $G - v$ and $G - \{v, w\}$ are cycles, which is clearly impossible. For the Merrifield-Simmons index, $G - v$ can be either isomorphic to C_{n-1} or to Y_{n-1} for equality, and $G - \{v, w\}$ can be isomorphic to either C_{n-2} or to Y_{n-2}. If $G - v$ is a cycle, then $G - \{v, w\}$ must be a path, and we cannot have equality. If $G - v$ is isomorphic to Y_{n-1}, then $G - \{v, w\}$ will only be connected and unicyclic if w is the unique pendant vertex, in which case $G - \{v, w\}$ is indeed isomorphic to Y_{n-2}. But then G has to be isomorphic to Y_n, which completes the proof. $\square$

Similar results to those obtained for unicyclic graphs in this section are also known for bicyclic and tricyclic graphs (including various versions with additional restrictions), see the survey [105] and the references therein. However, more general results for connected graphs with a given number of vertices and edges are only partially available. The problems of maximizing the Merrifield-Simmons index and minimizing the Hosoya index are somewhat simpler in this regard. The following general theorems can be found in [85, 98, 131]. Note that they incorporate parts of Theorems 4.3.2 and 4.3.3 as well as Theorem 4.6.2.

Theorem 4.6.4 *For every connected graph G with n vertices and m edges, where $n - 1 \leq m \leq 2n - 3$, we have the inequality*

$$Z(G) \geq mn - n^2 + 4n - 2m - 2,$$

and equality holds if G is (isomorphic to) the graph $S_{n,m}$ that is constructed as follows: let the vertices be $x, y,$ and $u_1, u_2, \ldots, u_{n-2}$. Now add an edge between x and y, connect $u_1, u_2, \ldots, u_{n-2}$ to y by an edge, and connect $u_1, u_2, \ldots, u_{m-n+1}$ to x by an edge (see Figure 4.11 for an example). If $m \neq n + 2$, this is the only graph (up to isomorphism) for which equality holds. For $m = n + 2$, equality holds for exactly one more graph, obtained by attaching $n - 4$ pendant vertices to a complete graph K_4.

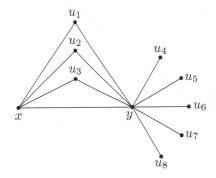

FIGURE 4.11
The graph $S_{n,m}$ in Theorem 4.6.4 and Theorem 4.6.5 (for $n = 10$ and $m = 12$).

Theorem 4.6.5 *For every connected graph G with n vertices and m edges, where $n - 1 \le m \le 2n - 3$, we have the inequality*

$$\sigma(G) \le 2^{n-2} + 2^{2n-m-3} + 1, \qquad (4.3)$$

and equality holds if G is the graph $S_{n,m}$ described in Theorem 4.6.4. If $m \ne 2n - 4$, this is the only graph (up to isomorphism) for which equality holds. For $m = 2n - 4$, equality holds for exactly one more graph, namely the complete bipartite graph $K_{2,n-2}$ (two vertices v, w are connected to all other $n - 2$ vertices by an edge, and no further edges).

In order to prove Theorem 4.6.4, we will need the following lemma:

Lemma 4.6.1 *Let G be a connected graph that is not a star, with the property that all its cut edges are pendant edges incident to a single vertex w. Then there exists an edge uv such that both $G - uv$ and $G - \{u, v\}$ are still connected.*

Proof:

Let T be a spanning tree of G. If there are two leaves in T that are connected by an edge in G, then we can choose these two leaves as u and v: $G - uv$ and $G - \{u, v\}$ are still connected, since T and $T - \{u, v\}$ are spanning trees for these two graphs. Thus we can assume that this is not the case. If T is a star, then it follows that there are no edges in G that are not already edges in T, which means that G is itself a star. Since this is also not the case by our assumptions, T cannot be a star. Now choose u to be a leaf in T whose distance to w is maximal (thus in particular greater than 1, since T is not a star), and let v be its unique neighbor in T. We claim that these two vertices satisfy the desired conditions. First note that uv is not a cut edge, since all those are assumed to be incident with w. Hence $G - uv$ is connected. Next note that all but one of the neighbors of v in T are leaves (the exception being the neighbor that lies on the path from w to v). If there was another non-leaf neighbor u', then we could extend the path from w to u' further to reach a leaf whose distance from w is greater than u's distance from w, which would contradict our choice of u. The leaf neighbors of v are not pendant vertices in G (since all pendant vertices are adjacent to w), and no two of them are adjacent (since we are assuming that no two leaves of T are adjacent). Hence each of them must have at least one neighbor in $G - \{u, v\}$, which means that we can extend $T - \{u, v\}$ to a spanning tree of $G - \{u, v\}$. This implies in particular that $G - \{u, v\}$ is connected, which completes our proof. $\qquad\square$

Proof of Theorem 4.6.4:

For $m = n - 1$, the graph is a tree, and the statement reduces to the upper bound in Theorem 4.3.3 (note also that the graph $S_{n,n-1}$ is indeed a star, where y is the center). Now we continue by induction on m. Let G be a graph with n vertices and m edges ($n \leq m \leq 2n - 3$) for which the Hosoya index attains its minimum. Suppose first that there is a cut edge vw that is not a pendant edge. Let A_1 and A_2 be the two components of $G - vw$ containing v and w, respectively. By Lemma 4.4.4, we can either move A_1 to w or A_2 to v (which actually amounts to the same) to decrease the Hosoya index, while the number of vertices and edges remains unchanged. This contradicts our choice of G, so all cut edges (if there are any) must be pendant edges. Now assume that there are two vertices v and w that are not pendant vertices, but adjacent to pendant vertices. We can use Lemma 4.4.4 again to show that moving pendant edges from v to w or from w to v decreases the Hosoya index, which is another contradiction to our choice of G. Thus all cut edges (if there are any) are pendant edges adjacent to a single vertex w. Moreover, since $m \geq n$, G is not a star, so Lemma 4.6.1 applies to G.

Thus we can pick an edge uv of G such that $G - uv$ and $G - \{u, v\}$ are connected. We can apply the induction hypothesis to $G - uv$ (which has $m - 1$ edges) to give us

$$Z(G - uv) \leq (m - 1)n - n^2 + 4n - 2(m - 1) - 2 = mn - n^2 + 3n - 2m.$$

The graph $G - \{u, v\}$ is connected and has $n - 2$ vertices, so we have $Z(G - \{u, v\}) \geq Z(S_{n-2}) = n - 2$. Combining the two bounds with Lemma 4.2.3 yields

$$Z(G) = Z(G - uv) + Z(G - \{u, v\}) \geq (mn - n^2 + 3n - 2m) + (n - 2)$$
$$= mn - n^2 + 4n - 2m - 2,$$

which completes our induction.

It remains to discuss the cases of equality: for equality to hold, $G - uv$ needs to be isomorphic to $S_{n,m-1}$ (except when $m = n + 3$, in which case there is one more possibility), and $G - \{u, v\}$ needs to be a star. Conversely, we need to add an edge uv between two vertices u and v of $S_{n,m-1}$ chosen in such a way that removing u and v from $S_{n,m-1}$ yields a star. It is not difficult to verify that the only possibility to do this is to add an edge between x and one of the pendant vertices of $S_{n,m-1}$, yielding a graph isomorphic to $S_{n,m}$, except when $m = n + 2$: in this case, another possibility is to add an edge between u_1 and u_2. However, it is not possible to extend the resulting graph (consisting of a complete graph K_4 and $n - 4$ pendant edges, as described) by another edge that satisfies the condition, so $m = n + 2$ is indeed the only case where there is more than one graph for which equality holds. $\qquad\square$

For the proof of Theorem 4.6.5, we follow the arguments given in [102], which are rather different from the proof of Theorem 4.6.4 even though the statements are analogous.

Proof of Theorem 4.6.5:

We gradually rule out certain substructures. First, suppose that the graph G contains a path of length 5 as a (not necessarily induced) subgraph. This is only possible if $n \geq 6$. We can extend this path to a spanning tree T whose diameter must be at least 5. By Theorem 4.5.1, we have $\sigma(T) \leq 2^{n-2} + 5$. Moreover, for each edge that needs to be added to T to obtain G, the Merrifield-Simmons index decreases at least by 1. Thus $\sigma(G) \leq \sigma(T) - (m - n + 1) \leq 2^{n-2} - (m - n - 4)$. We can use the well-known inequality $2^k > k$, which is valid for all integers k, and the fact that $n \geq 6$, to obtain

$$\sigma(G) \leq 2^{n-2} - (m - n - 4) = 2^{n-2} + (n - m + 3) + 1$$
$$< 2^{n-2} + 2^{n-m+3} + 1 \leq 2^{n-2} + 2^{2n-m-3} + 1,$$

which means that (4.3) even holds with strict inequality.

So in the following, we can assume that there are no paths of length 5 in G, thus also no cycles of length 6 or more (since such a cycle would contain a path of length 5). Next, suppose that a cycle of length 5 occurs as a subgraph of G. If there is at least one more vertex adjacent to a vertex of this cycle, then there is a path of length 5, which is impossible. Since G is connected, it follows that the cycle vertices are the only vertices of G, so $n = 5$. Since $\sigma(C_5) = L_5 = 11$, the same argument as before gives us

$$\sigma(G) \leq \sigma(C_5) - (m - 5) = 16 - m.$$

It is easily verified that we have $16 - m < 2^{n-2} + 2^{2n-m-3} + 1 = 9 + 2^{7-m}$ in all possible cases, so again (4.3) holds with strict inequality.

Thus all cycles in G have either length 3 or 4. Assume that there is no cycle of length 4. If there are two disjoint cycles of length 3, then there must be a path connecting the two. This path, together with two edges from each of the cycles, yields a path of length 5, which has already been ruled out. So any two cycles share at least one vertex. If there are two cycles that even share an edge, then they form a cycle of length 4, which contradicts our assumption. So the only possibility remaining is that any two cycles share exactly one vertex.

Let u_1u_2w and v_1v_2w be two such cycles. If there is another vertex x adjacent to u_1, then we have a path of length 5, namely $xu_1u_2wv_1v_2$. Since this has been ruled out, there is no such vertex x, and analogously also no further vertex adjacent to u_2, v_1 or v_2. Next suppose that there are two vertices $x, y \notin \{u_1, u_2, v_1, v_2, w\}$ that are adjacent to each other. The subgraph formed by the two cycles has Merrifield-Simmons index 10, the subgraph formed by x and y has Merrifield-Simmons index 3. Thus if we remove all edges except for the two cycles and the edge xy, we get a subgraph G' of G with $\sigma(G') = 30 \cdot 2^{n-7}$, and it follows that

$$\sigma(G) \leq \sigma(G') = 30 \cdot 2^{n-7} < 2^{n-2} < 2^{n-2} + 2^{2n-m-3} + 1.$$

Once again, strict inequality holds. The only remaining possibility is that G consists of the two cycles u_1u_2w and v_1v_2w and a number of pendant vertices attached to w. This graph has $m = n + 1$ edges, and we can apply the second item of Lemma 4.2.2 to w to find

$$\sigma(G) = 9 \cdot 2^{n-5} + 1 < 10 \cdot 2^{n-5} + 1 = 2^{n-2} + 2^{2n-m-3} + 1,$$

and we are done again.

Now we are left with three cases: there is no cycle at all, only one cycle of length 3 (and no other cycle), or at least one cycle of length 4. The former two cases are covered by Theorem 4.3.2 and Theorem 4.6.2, respectively. So let $u_1u_2u_3u_4$ be a cycle of length 4. If there are two distinct vertices v and w that do not lie on the cycle and are adjacent to u_1 and u_2, respectively, then there is a path of length 5 (namely $vu_1u_4u_3u_2w$), which has been ruled out. The same argument applies to other pairs of vertices that are adjacent on the cycle. Two vertices v and w outside of the cycle cannot be adjacent to each other either: there must be a path connecting them to the cycle, and the combination of the edge vw, the connecting path and three cycle edges would give us a path of length 5 or more. So vertices that do not lie on the cycle can only be adjacent to vertices on the cycle, and there are at most two cycle vertices adjacent to other vertices, which cannot be adjacent on the cycle. Without loss of generality, let these two be u_1 and u_3. If there are pendant vertices adjacent to both u_1 and u_3, then we can either move pendant vertices from u_1 to u_3 or from u_3 to u_1 to increase the Merrifield-Simmons index (by Lemma 4.4.4). Thus it suffices to consider the case where all pendant vertices are adjacent to u_1. All other vertices that do not lie on the cycle are adjacent to both u_1 and u_3. If there is an edge between u_2 and u_4, then there cannot be any vertex outside of the cycle that is adjacent to u_1 and u_3: if v was such a vertex, then $vu_1u_2u_4u_3v$ would be a cycle of length 5. So we are left with the following scenarios (based on the distinction whether or not the edges u_1u_3 and u_2u_4 are present in the graph):

- Both u_1u_3 and u_2u_4 are edges in G. Then all other vertices must be pendant vertices adjacent to v_1. We have $m = n + 2$, and we easily calculate

$$\sigma(G) = 2^{n-2} + 1 < 2^{n-2} + 2^{n-5} + 1 = 2^{n-2} + 2^{2n-m-3} + 1.$$

- $u_1 u_3$ is not an edge in G, but $u_2 u_4$ is. Again, all other vertices must be pendant vertices adjacent to v_1. Now $m = n + 1$, and we obtain

$$\sigma(G) = 2^{n-2} + 2 \le 2^{n-2} + 2^{n-4} + 1 = 2^{n-2} + 2^{2n-m-3} + 1,$$

  with equality only if $n = 4$ (in which case the graph G is indeed isomorphic to $S_{4,5}$).

- $u_2 u_4$ is not an edge in G, but $u_1 u_3$ is. In this case, we find that G is exactly the graph $S_{n,m}$, and we can use Lemma 4.2.2 to calculate $\sigma(G)$:

$$\sigma(G) = \sigma(S_{n,m}) = 2^{n-2} + 2^{2n-m-3} + 1.$$

- $u_1 u_3$ and $u_2 u_4$ are not edges in G. In this case, the number of edges is at most $4 + 2(n - 4) = 2n - 4$. Again, we use Lemma 4.2.2 to calculate the Merrifield-Simmons index directly:

$$\sigma(G) = 2^{n-2} + 2^{2n-m-4} + 2 \le 2^{n-2} + 2^{2n-m-3} + 1,$$

  and equality holds only if $m = 2n - 4$, in which case G is the complete bipartite graph $K_{2,n-2}$.

  $\square$

4.7   Independence polynomial and matching polynomial

There are many different polynomials that can be associated with a graph. The *independence polynomial* and the *matching polynomial* are two prominent examples, and further examples will follow in the next chapter.

Let $i(G, k)$ be the number of independent sets of G consisting of exactly k vertices (this includes the case $k = 0$, where $i(G, 0) = 1$ for all graphs since the empty set is always independent). The independence polynomial $I(G; x)$ of G is defined as

$$I(G; x) = \sum_{k \ge 0} i(G, k) x^k.$$

An alternative way of expressing this definition is as follows: if $\mathcal{I}(G)$ denotes the set of all independent sets of G, then

$$I(G; x) = \sum_{A \in \mathcal{I}(G)} x^{|A|}.$$

It is clear that there are $i(G, k)$ terms equal to x^k in this sum, so the two definitions are indeed equivalent. The connection to the Merrifield-Simmons index is immediate: we have

$$\sigma(G) = I(G; 1).$$

The definition of the matching polynomial is slightly different, but similar. Let $m(G, k)$ be the number of matchings of G consisting of k edges (note again that $m(G, 0) = 1$ for all graphs G), and set

$$M(G; x) = \sum_{k \geq 0} (-1)^k m(G, k) x^{n-2k},$$

where n is the number of vertices of G. The exponent $n - 2k$ is always non-negative, since $2k$ is the number of vertices covered by a matching of cardinality k, which cannot exceed n. This definition might seem unintuitive at first, but we will see some reasons for it later in this section and in particular in the next chapter. The somewhat more natural-looking matching generating polynomial

$$\mu(G; x) = \sum_{k \geq 0} m(G, k) x^k$$

is connected to the Hosoya index in the same way the independence polynomial is connected to the Merrifield-Simmons index, namely by $\mu(G; 1) = Z(G)$. Note that $M(G; x)$ and $\mu(G; x)$, although slightly different, are closely related: we have

$$M(G; x) = x^n \mu(G; -x^{-2}). \tag{4.4}$$

As before, there is an alternative way to express the definitions: let $\mathcal{M}(G)$ be the set of all matchings of G. Then we have

$$M(G; x) = \sum_{B \in \mathcal{M}(G)} (-1)^{|B|} x^{n-2|B|}$$

and

$$\mu(G; x) = \sum_{B \in \mathcal{M}(G)} x^{|B|}.$$

Let us go back to the example of Figure 4.1, which we used to illustrate the Merrifield-Simmons index and the Hosoya index. This graph has six independent sets: the empty set, four single-vertex sets, and one independent set of two vertices. Thus

$$I(G; x) = x^2 + 4x + 1.$$

Its eight matchings are: the empty set, five single-edge sets, and two matchings consisting of two edges. So we have

$$M(G; x) = x^4 - 5x^2 + 2 \qquad \text{and} \qquad \mu(G; x) = 2x^2 + 5x + 1.$$

Now we briefly discuss some of the properties of the independence polynomial and the matching polynomial (and matching generating polynomial). Recall from Proposition 4.2.1 that the Hosoya index and the Merrifield-Simmons index can be connected via the concept of a line graph. This extends to the respective polynomials, and the proof is equally straightforward:

Proposition 4.7.1 *For every graph G, we have $\mu(G; x) = I(\mathcal{L}(G); x)$.*

The simple recursive relations of Lemma 4.2.2 and Lemma 4.2.3 proved useful throughout this chapter. Analogues of these relations also hold for the independence polynomial and the matching polynomial, and they are stated in the following:

Lemma 4.7.1 • *If $G_1, G_2, \ldots, G_k$ are the connected components of a graph G, then we have*

$$I(G; x) = \prod_{j=1}^{k} I(G_j; x).$$

• *For every vertex v of G, we have*

$$I(G; x) = I(G - v; x) + xI(G - N[v]; x).$$

• *For every edge e of G whose ends are v and w, we have*

$$I(G; x) = I(G - e; x) - x^2 I(G - (N[v] \cup N[w]); x).$$

Note that all these formulas reduce to Lemma 4.2.2 if we plug in $x = 1$.

Proof:

• We can make use of the fact that $\mathcal{I}(G) = \mathcal{I}(G_1) \times \mathcal{I}(G_2) \times \cdots \times \mathcal{I}(G_k)$: every independent set of G induces independent sets in all components, and conversely, the union of independent sets in all components is an independent set in G. Thus

$$I(G; x) = \sum_{A \in \mathcal{I}(G)} x^{|A|} = \sum_{A_1 \in \mathcal{I}(G_1)} \sum_{A_2 \in \mathcal{I}(G_2)} \cdots \sum_{A_k \in \mathcal{I}(G_k)} x^{|A_1| + |A_2| + \cdots + |A_k|}$$

$$= \sum_{A_1 \in \mathcal{I}(G_1)} x^{|A_1|} \sum_{A_2 \in \mathcal{I}(G_2)} x^{|A_2|} \cdots \sum_{A_k \in \mathcal{I}(G_k)} x^{|A_k|}$$

$$= I(G_1; x) I(G_2; x) \cdots I(G_k; x).$$

• As in the proof of Lemma 4.2.2, we make the observation that an independent set of G either does not contain v, or it contains v but none of v's neighbors. Thus

$$I(G; x) = \sum_{A \in \mathcal{I}(G)} x^{|A|} = \sum_{\substack{A \in \mathcal{I}(G) \\ v \notin A}} x^{|A|} + \sum_{\substack{A \in \mathcal{I}(G) \\ v \in A}} x^{|A|}$$

$$= \sum_{A \in \mathcal{I}(G-v)} x^{|A|} + \sum_{A \in \mathcal{I}(G-N[v])} x^{|A \cup \{v\}|}$$

$$= I(G - v; x) + xI(G - N[v]; x).$$

- For the final identity, we note that independent sets of $G - e$ are either independent sets of G or contain both v and w (but none of their neighbors). Consequently, we have

$$I(G;x) = \sum_{A \in \mathcal{I}(G)} x^{|A|} = \sum_{A \in \mathcal{I}(G-e)} x^{|A|} - \sum_{\substack{A \in \mathcal{I}(G-e) \\ v,w \in A}} x^{|A|}$$

$$= \sum_{A \in \mathcal{I}(G-e)} x^{|A|} - \sum_{A \in \mathcal{I}(G-(N[v] \cup N[w]))} x^{|A \cup \{v,w\}|}$$

$$= I(G - e;x) - x^2 I(G - (N[v] \cup N[w]);x).$$

This completes the proof.

$\square$

In the same vein, we have the following lemma on the matching polynomial:

Lemma 4.7.2 • *If $G_1, G_2, \ldots, G_k$ are the connected components of a graph G, then we have*

$$M(G;x) = \prod_{j=1}^{k} M(G_j;x) \qquad and \qquad \mu(G;x) = \prod_{j=1}^{k} \mu(G_j;x).$$

- *For every vertex v of G, we have*

$$M(G;x) = xM(G - v;x) - \sum_{w \in N(v)} M(G - \{v,w\};x)$$

and

$$\mu(G;x) = \mu(G - v;x) + x \sum_{w \in N(v)} \mu(G - \{v,w\};x).$$

- *For every edge e of G whose ends are v and w, we have*

$$M(G;x) = M(G - e;x) - M(G - \{v,w\};x)$$

and

$$\mu(G;x) = \mu(G - e;x) + x\mu(G - \{v,w\};x).$$

Proof:

It suffices to prove all statements for $\mu(G;x)$, the corresponding formulas for $M(G;x)$ follow directly from (4.4). The proof of the first formula is completely analogous to the first part of Lemma 4.7.1, so we focus on the second and third formula:

- As in the proof of Lemma 4.2.3, we observe that a matching of G either does not cover v at all, or it contains exactly one of the edges vw where $w \in N(v)$. Thus we have

$$\mu(G; x) = \sum_{A \in \mathcal{M}(G)} x^{|A|}$$

$$= \sum_{A \in \mathcal{M}(G-v)} x^{|A|} + \sum_{\substack{w \in N(v)}} \sum_{\substack{A \in \mathcal{M}(G) \\ vw \in A}} x^{|A|}$$

$$= \sum_{A \in \mathcal{M}(G-v)} x^{|A|} + \sum_{\substack{w \in N(v)}} \sum_{\substack{A \in \mathcal{M}(G-\{v,w\})}} x^{|A \cup \{vw\}|}$$

$$= \mu(G - v; x) + x\mu(G - \{v, w\}; x).$$

- Similarly, we can partition the set of matchings of G into those that contain $e = vw$ and those that do not:

$$\mu(G; x) = \sum_{A \in \mathcal{M}(G)} x^{|A|} = \sum_{\substack{A \in \mathcal{M}(G) \\ e \notin A}} x^{|A|} + \sum_{\substack{A \in \mathcal{M}(G) \\ e \in A}} x^{|A|}$$

$$= \sum_{A \in \mathcal{M}(G-e)} x^{|A|} + \sum_{\substack{A \in \mathcal{M}(G-\{v,w\})}} x^{|A \cup \{e\}|}$$

$$= \mu(G - e; x) + x\mu(G - \{v, w\}; x),$$

and the proof is complete.

$\square$

A famous result due to Heilmann and Lieb [48], who studied the matching polynomial in the context of the monomer-dimer model in statistical physics, states that the zeros of the matching polynomial of a graph are always real. In fact, an even stronger statement holds, which is known as the interlacing property. We will obtain this result by first proving a slightly more general version that also involves edge weights. Suppose that each edge $e = vw$ of a graph is assigned a weight w_{vw}. We define the weight $\omega(B)$ of a matching B to be the product of the weights of all edges occurring in it. The matching polynomial associated with a weighted graph G is defined as

$$M(G; x) = \sum_{B \in \mathcal{M}(G)} (-1)^{|B|} \omega(B) x^{n-2|B|}.$$

Observe that the unweighted version is obtained simply by setting all weights equal to 1. It will be important for us that the second formula of Lemma 4.7.2 still holds with a small modification to account for the weights:

$$M(G; x) = xM(G - v; x) - \sum_{w \in N(v)} w_{vw} M(G - \{v, w\}; x). \qquad (4.5)$$

The proof is analogous to the unweighted version.

Theorem 4.7.1 *Let G be a graph with n vertices, and let G_1 be an induced subgraph with $n - 1$ vertices, i.e., a graph obtained by removing exactly one vertex from G. All the zeros of $M(G; x)$ and $M(G_1; x)$ are real. Moreover, if we let $\alpha_1 \leq \alpha_2 \leq \cdots \leq \alpha_n$ be the zeros of $M(G; x)$ in non-decreasing order, and let $\beta_1 \leq \beta_2 \leq \cdots \leq \beta_{n-1}$ be the zeros of $M(G_1; x)$ in non-decreasing order, then we have*

$$\alpha_1 \leq \beta_1 \leq \alpha_2 \leq \beta_2 \leq \cdots \leq \beta_{n-2} \leq \alpha_{n-1} \leq \beta_{n-1} \leq \alpha_n.$$

Proof:

We will first prove a version of the theorem for weighted complete graphs by induction on the number of vertices. The desired statement will then follow by a limiting argument. Let K be a weighted complete graph with n vertices and positive edge weights (i.e., for every pair v, w of vertices, we have $\omega_{vw} > 0$). Then the following hold:

- The matching polynomial of K has distinct real zeros $\alpha_1 < \alpha_2 < \cdots < \alpha_n$.

- If K' is obtained from K by removing one of the vertices, then the zeros $\beta_1, \beta_2, \ldots, \beta_{n-1}$ of the matching polynomial of K interlace with the zeros of K, i.e.,

$$\alpha_1 < \beta_1 < \alpha_2 < \beta_2 < \cdots < \beta_{n-2} < \alpha_{n-1} < \beta_{n-1} < \alpha_n.$$

For $n = 2$, the statement is easy to verify: let v and w be the two vertices and ω_{vw} the only edge weight. The matching polynomial of K is

$$x^2 - \omega_{vw},$$

which has the real zeros $\pm\sqrt{\omega_{vw}}$. The matching polynomial of K' (which only has one vertex, but no edges) is x, whose only zero is 0. Clearly, the interlacing property is satisfied. For the induction step, we use the recursive formula (4.5): let v be one of the vertices of K, and let K' be obtained by removing v from K. Then we have

$$M(K; x) = xM(K'; x) - \sum_{w \neq v} \omega_{vw} M(K - \{v, w\}; x).$$

Note that $K - \{v, w\}$ can be obtained from K' by removing the vertex w, so we can apply the induction hypothesis to K' and $K - \{v, w\}$ for every w. In particular, we know that the zeros of K' are all real and distinct, so that we can arrange them in increasing order: $\beta_1 < \beta_2 < \cdots < \beta_{n-1}$.

Now consider one of the graphs $K - \{v, w\}$: let the zeros of its matching polynomial be $\gamma_1, \gamma_2, \ldots, \gamma_{n-2}$, so that

$$\beta_1 < \gamma_1 < \beta_2 < \gamma_2 < \cdots < \gamma_{n-3} < \beta_{n-2} < \gamma_{n-2} < \beta_{n-1}$$

by the induction hypothesis.

Since the zeros of the polynomial $P(x) = M(K - \{v, w\}; x)$ are all distinct, the sign of the polynomial changes at each of the zeros. Moreover, by definition of the matching polynomial, the leading term of $P(x)$ is x^{n-2}. This implies that $P(x) > 0$ for $x > \gamma_{n-2}$, and the signs alternate on intervals of the form (γ_{i-1}, γ_i): we have $(-1)^{n+i-1}P(x) > 0$ for $x \in (\gamma_{i-1}, \gamma_i)$ and $(-1)^n P(x) > 0$ for $x < \gamma_1$. In particular, we must have $(-1)^{n+i-1}P(\beta_i) = (-1)^{n+i-1}M(K - \{v, w\}; \beta_i) > 0$. Since this holds for all vertices w and the weights ω_{vw} were assumed to be positive, it follows that

$$(-1)^{n+i}M(K; \beta_i) = (-1)^{n+i}\left(xM(K'; \beta_i) - \sum_{w \neq v} \omega_{vw}M(K - \{v, w\}; \beta_i)\right)$$

$$= (-1)^{n+i-1} \sum_{w \neq v} \omega_{vw}M(K - \{v, w\}; \beta_i)$$

$$> 0.$$

Therefore, the signs of the values of $M(K; x)$ at the zeros of $M(K'; x)$ alternate: $M(K; \beta_{i-1})$ and $M(K; \beta_i)$ have different signs. By the intermediate value theorem, there must be a zero of $M(K; x)$ in the interval (β_{i-1}, β_i). This gives us $n - 2$ distinct roots of $M(K; x)$ that interlace with the roots of $M(K; x)$. The remaining two roots are found by a similar argument: we have $M(K; \beta_{n-1}) < 0$, but since the leading term of $M(K; x)$ is x^n, we have $\lim_{x \to \infty} M(K; x) = +\infty$. Thus there must also be a zero of $M(K; x)$ that is greater than β_{n-1}, and a similar argument shows that there is also a zero of $M(K; x)$ less than β_1. The desired statement follows.

Now we want to generalize to the situation that the graph is no longer complete: we can interpret a given graph G as a weighted complete graph with weights given by $\omega_{vw} = 1$ if vw is an edge in G and $\omega_{vw} = 0$ otherwise: it is clear that a matching of the complete graph is assigned weight 1 if it is a matching in G, and 0 otherwise. However, the weights are not strictly positive in this setting. So instead, we define weights by $\omega_{vw} = 1$ if vw is an edge and $\omega_{vw} = \epsilon > 0$ otherwise. The coefficients of the matching polynomial of the resulting graph are continuous functions of ϵ. Moreover, it is well known that the zeros of a monic polynomial are continuous functions of the coefficients. Therefore, the zeros of the matching polynomial of the weighted complete graph defined by these weights converge to the zeros of $M(G; x)$ as $\epsilon \to 0$. Thus Theorem 4.7.1 follows from the statement on complete graphs with positive weights (note that strict inequalities may become non-strict in the limit). $\square$

Corollary 4.7.1 *For every graph G, the zeros of the matching generating polynomial $\mu(G; x)$ are real and negative.*

Proof:

Recall that

$$M(G; x) = x^n \mu(G; -x^{-2}),$$

so the zeros of $\mu(G;x)$ are of the form $-\alpha^{-2}$, where α is a zero of $M(G;x)$. Clearly, all these zeros are real and negative. Note that 0 is not a zero of $\mu(G;x)$, since $\mu(G;x) = m(G,0) = 1$ for every graph G. ☐

Let us remark that Theorem 4.7.1 holds for weighted graphs as well, provided that all weights are non-negative (by an analogous argument). Notably, the zeros of the independence polynomial of a graph are not always real. As a counterexample, consider the star with four vertices. It has one independent set of three vertices (consisting of the three leaves), three independent sets of two vertices (any two leaves), four single-vertex independent sets, and of course the empty set as another independent set. Thus

$$I(S_4;x) = x^3 + 3x^2 + 4x + 1,$$

and this polynomial has only one real zero (and two complex conjugate zeros). It is known, however, that the zeros are real if the graph is claw-free (no induced subgraph is isomorphic to the claw, i.e., the four-vertex star); this was proven by Chudnovsky and Seymour [16].

It was mentioned earlier in this section that the Merrifield-Simmons index and the Hosoya index are closely related to the independence polynomial and the matching generating polynomial via the identities

$$\sigma(G) = I(G;1) \qquad \text{and} \qquad Z(G) = \mu(G;1).$$

Very frequently, graphs that have the maximum or minimum total number of independent sets or matchings in a certain class of graphs (thus giving the maximum or minimum value of $I(G;1)$ or $\mu(G;1)$) are even coefficientwise optimal in that they maximize or minimize every coefficient of the respective polynomial.

The idea of introducing an order structure on graphs based on the coefficients of the independence polynomial or matching polynomial goes back to the papers [34, 35] by Gutman. Let us define the quasi-order $\preceq_i$ in the following way: for graphs G and H (usually with the same number of vertices), we write $G \preceq_i H$ if $i(G,k) \leq i(H,k)$ holds for all k. If at least one of these inequalities holds with strict inequality, then we write $G \prec_i H$. The relations $\preceq_m$ and $\prec_m$ are defined in an analogous way, with $i(\cdot,k)$ replaced by $m(\cdot,k)$.

It is clear that $G \preceq_i H$ implies $\sigma(G) \leq \sigma(H)$, and $G \prec_i H$ implies $\sigma(G) < \sigma(H)$. Likewise, $G \preceq_m H$ implies $Z(G) \leq Z(H)$, and $G \prec_m H$ implies $Z(G) < Z(H)$. The following theorem, which extends Theorem 4.3.1, is essentially trivial (it follows from the same monotonicity argument):

Theorem 4.7.2 *For every graph G with n vertices, we have*

$$K_n \preceq_i G \qquad and \qquad G \preceq_i E_n,$$

where K_n and E_n are the complete and edgeless graph, respectively. Similarly,

$$E_n \preceq_m G \qquad and \qquad G \preceq_m K_n.$$

For trees, we also have the following extensions of Theorem 4.3.2 and Theorem 4.3.3, which can in fact be proved along the same lines. This theorem will become particularly useful in the following chapter in our discussion of the graph energy.

Theorem 4.7.3 *For every tree T with n vertices, the following statements hold:*

- *$T \preceq_i S_n$, and $T \prec_i S_n$ unless T is the star. Likewise, $T \succeq_i P_n$, and $T \succ_i P_n$ unless T is the path.*

- *$T \succeq_m S_n$, and $T \succ_m S_n$ unless T is the star. Likewise, $T \preceq_m P_n$, and $T \prec_m P_n$ unless T is the path.*

Proof:

- We start with independent sets, our goal being to show that

$$i(P_n, k) \leq i(T, k) \leq i(S_n, k), \tag{4.6}$$

and that equality in the lower bound for all k is only possible if T is the path, while equality in the upper bound for all k is only possible if T is the star. The argument that gave us the second item of Lemma 4.2.2 also shows that

$$i(G, k) = i(G - v, k) + i(G - N[v], k - 1)$$

for every graph G and every vertex v of G. We will apply this recursive formula to prove both statements by induction. Note first that both are trivial for $n = 1$ and $n = 2$ (as there is only one tree in these cases). For the induction step, let v be a leaf of T, and let w be its unique neighbor. By the induction hypothesis, we have

$$i(P_{n-1}, k) \leq i(T - v, k) \leq i(S_{n-1}, k),$$

as well as

$$i(P_{n-2}, k - 1) \leq i(T - \{v, w\}, k - 1) \leq i(E_{n-2}, k - 1),$$

the latter by Theorem 4.7.2. Adding the two inequalities, we arrive at (4.6). If equality holds for all k in the lower bound, then we must in particular have $\sigma(P_n) = \sigma(T)$, so Theorem 4.3.2 shows that T must be a path. Likewise, if equality holds for all k in the upper bound, then $\sigma(S_n) = \sigma(T)$, which implies that T is a star.

- The proof for matchings is completely analogous (Exercise 16). For the statement that $T \succeq_m S_n$, we can also argue as in the proof of Theorem 4.3.3: we have $m(T, 0) = m(S_n, 0) = 1$ and $m(T, 1) = m(S_n, 1) = n - 1$ for all trees T with n vertices, and $m(S_n, k) = 0$ for all $k > 1$. Moreover, $m(T, 2) > 0$ for every tree that is not a star, so that we even have $T \succ_m S_n$ in this case.

$\square$

Exercises

1. A ladder H_n consists of two n-vertex paths $v_1, v_2, \ldots, v_n$ and $w_1, w_2, \ldots, w_n$ and additional edges $v_1 w_1, v_2 w_2, \ldots, v_n w_n$. Determine formulas for $\sigma(H_n)$ and $Z(H_n)$.

2. A wheel W_n consists of a cycle with vertices $v_1, v_2, \ldots, v_n$ and an additional vertex w that is connected to all vertices of the cycle by an edge. Determine formulas for $\sigma(W_n)$ and $Z(W_n)$.

3. Let a tripod be a tree with precisely one vertex of degree 3 and no further vertices of degree 3 or more. Prove that $\sigma(T) + Z(T) = F_{n+3}$ holds for every tripod T with n vertices.

4. Prove Lemma 4.4.1 for the Hosoya index.

5. Prove the identities $F_{k+1} F_{h+1} + F_k F_h = F_{k+h+1}$ and $F_{k+2} F_{h+2} - F_k F_h = F_{k+h+2}$ by applying Lemma 4.2.2 to a path.

6. Verify that the expressions for $\sigma(G) - \sigma(G_v)$ and $\sigma(G) - \sigma(G_w)$ in the proof of Lemma 4.4.5 remain correct for $r \leq 3$, when $\sigma(P_{r-3})$ and other expressions are not well defined.

7. Determine the n-vertex tree with third-smallest/third-largest Merrifield-Simmons index/Hosoya index.

8. Complete the proof of Theorem 4.5.2 for the Hosoya index.

9. Show that Theorem 4.5.1 can be derived from Theorems 4.5.4 and 4.5.5.

10. Verify that the $\mathcal{M}$-tree $\mathcal{M}(\ell, 2, 2, \ldots, 2, 1, 1, \ldots, 1)$ is indeed a comet, as claimed at the end of Section 4.5.

11. Deduce Theorem 4.6.3 from Theorem 4.6.1.

12. A graph G is called a quasi-tree if it can be turned into a tree by removing a single vertex. Prove that the minimum of the Merrifield-Simmons index and the maximum of the Hosoya index among quasi-trees with n vertices are attained by the fan, which is the graph obtained from an $(n-1)$-vertex path by adding a vertex and connecting it to all vertices of the path by an edge.

13. Find an interpretation for the two expressions

$$\frac{d}{dx} \ln I(G; x) \Big|_{x=1} \quad \text{and} \quad \frac{d}{dx} \ln \mu(G; x) \Big|_{x=1}.$$

14. Find two non-isomorphic graphs with the same matching polynomial, and two non-isomorphic graphs with the same independence polynomial.

15. Find an example of two graphs G and H with the same number of vertices for which neither $G \preceq_i H$ nor $H \preceq_i G$ holds. Find an analogous example for $\preceq_m$.

16. Prove the second part of Theorem 4.7.3 by means of a similar induction argument as in the first part.

5

Graph spectra and the graph energy

5.1 Matrices associated with graphs

There are several natural matrices that can be associated with a graph. The best-known instance is probably the adjacency matrix $A(G)$ of a graph G. Fix an order of the vertices of G as $v_1, v_2, \ldots, v_n$. The entries a_{ij} of $A(G)$ are defined as follows:

$$a_{ij} - \begin{cases} 1 & \text{if } v_i \text{ and } v_j \text{ are adjacent,} \\ 0 & \text{otherwise.} \end{cases}$$

For the example in Figure 5.1 (the same example we used at the beginning of Section 4.1), the adjacency matrix looks as follows:

$$A(G) = \begin{bmatrix} 0 & 1 & 0 & 1 \\ 1 & 0 & 1 & 1 \\ 0 & 1 & 0 & 1 \\ 1 & 1 & 1 & 0 \end{bmatrix}.$$

We observe that the adjacency matrix of a (simple, undirected) graph is a symmetric 0-1-matrix, and all diagonal entries are 0s. It is also important to notice that the adjacency matrix is not unique: it depends on the order of vertices chosen. However, this dependence will actually prove immaterial in the following.

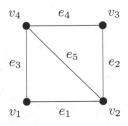

FIGURE 5.1
Example graph to illustrate the definition of adjacency matrix, incidence matrix, and (signless) Laplacian matrix.

The incidence matrix $B(G)$ associated with a graph is defined in a very similar way. In order to define it, we also fix an order for the edges: $e_1, e_2, \ldots, e_m$. The entries b_{ij} of $B(G)$ are defined as follows:

$$b_{ij} = \begin{cases} 1 & \text{if } v_i \text{ is one of the ends of } e_j, \\ 0 & \text{otherwise.} \end{cases}$$

Looking again at the example in Figure 5.1, the incidence matrix is

$$B(G) = \begin{bmatrix} 1 & 0 & 1 & 0 & 0 \\ 1 & 1 & 0 & 0 & 1 \\ 0 & 1 & 0 & 1 & 0 \\ 0 & 0 & 1 & 1 & 1 \end{bmatrix}.$$

The adjacency and incidence matrix are related by several interesting identities such as the following:

Proposition 5.1.1 *The incidence matrix of G and the adjacency matrix of the line graph (with respect to the same ordering) are related by*

$$A(\mathcal{L}(G)) = B(G)^T B(G) - 2I,$$

where I denotes the identity matrix.

Proof:

The entry in the i-th row, j-th column of $B(G)^T B(G)$ is the inner product of the i-th column and the j-th column of $B(G)$. Both columns only have two non-zero entries that are equal to 1. If $i \neq j$, then the two columns have either no common entry (if the corresponding edges do not share a vertex), or exactly one (if the corresponding edges share a vertex). In the former case, the inner product is 0, in the latter it is 1. In both cases, it agrees with the corresponding entry in $A(\mathcal{L}(G))$. If $i = j$, the two columns are identical and the inner product is 2. The final term $2I$ in the formula removes these entries so that they agree with the diagonal entries of $A(\mathcal{L}(G))$. This completes the proof. □

If the incidence matrix and its transpose are multiplied in a different order, we obtain another important matrix associated with a graph, namely the signless Laplacian. The signless Laplacian $S(G)$ has entries s_{ij} defined as follows:

$$s_{ij} = \begin{cases} \deg(v_i) & \text{if } i = j, \\ 1 & \text{if } i \neq j \text{ and } v_i \text{ and } v_j \text{ are adjacent}, \\ 0 & \text{otherwise.} \end{cases}$$

If we let $D(G)$ (the degree matrix) be the diagonal matrix whose entries are the degrees $\deg(v_i)$ of the vertices, then we can express $S(G)$ as

$$S(G) = D(G) + A(G).$$

We have the following identity that we alluded to earlier:

Proposition 5.1.2 *The incidence matrix and the signless Laplacian of G (with respect to the same vertex ordering) are related by*
$$S(G) = B(G)B(G)^T.$$

Proof:

In analogy to the proof of Proposition 5.1.1, we note that the entry in the i-th row, j-th column of $B(G)B(G)^T$ is the inner product of the i-th row and the j-th row of $B(G)$. Since both are 0-1-vectors, the inner product is equal to the number of positions where both have an entry 1. If $i = j$, then the rows are identical and contain $\deg(v_i)$ 1s (one for each edge incident with v_i), so the inner product is $\deg(v_i)$. Otherwise, the two rows only have a common entry 1 in a column corresponding to an edge of which both are ends. Thus the inner product is 1 if v_i and v_j are adjacent, and 0 otherwise. We conclude that the entries of $B(G)B(G)^T$ agree with those of $S(G)$. □

The matrix
$$L(G) = D(G) - A(G)$$
is called the Laplacian of G (more precisely, the combinatorial Laplacian, to distinguish it from its probabilistic counterpart). Except for the sign, its entries are exactly the same as those of $S(G)$. An analogous formula to Proposition 5.1.2 holds, where $B(G)$ needs to be replaced by a signed incidence matrix: fix an orientation of G by picking (arbitrarily) a head $h(e)$ and a tail $t(e)$ for each edge e. Then define the oriented incidence matrix $C(G)$ by its entries c_{ij}:
$$c_{ij} = \begin{cases} 1 & \text{if } v_i \text{ is the head of } e_j, \\ -1 & \text{if } v_i \text{ is the tail of } e_j, \\ 0 & \text{otherwise.} \end{cases}$$

Proposition 5.1.3 *The oriented incidence matrix $C(G)$ as defined above and the Laplacian of G are related by*
$$L(G) = C(G)C(G)^T.$$

Proof:

Analogous to Proposition 5.1.2. □

To illustrate the concept of the Laplacian and the signless Laplacian, we refer to Figure 5.1 once again. For the graph in this figure, the matrices $L(G)$ and $S(G)$ are given by
$$L(G) = \begin{bmatrix} 2 & -1 & 0 & -1 \\ -1 & 3 & -1 & -1 \\ 0 & -1 & 2 & -1 \\ -1 & -1 & -1 & 3 \end{bmatrix} \quad \text{and} \quad S(G) = \begin{bmatrix} 2 & 1 & 0 & 1 \\ 1 & 3 & 1 & 1 \\ 0 & 1 & 2 & 1 \\ 1 & 1 & 1 & 3 \end{bmatrix}.$$

Before we consider spectral properties of the matrices defined in this section, let us discuss a relation between the adjacency matrix and walks in graphs: by a walk of length r, we mean a sequence $u_0, u_1, u_2, \ldots, u_r$ of (not necessarily distinct) vertices such that u_i and u_{i+1} are adjacent for all i. The adjacency matrix can be used to count these walks:

Proposition 5.1.4 *For two vertices v_i and v_j of a graph G, let $w_{ij}^{(r)}$ be the number of walks of length r whose first vertex is v_i and whose last vertex is v_j. Then $w_{ij}^{(r)}$ is equal to the entry in the i-th row, j-th column of $A(G)^r$.*

Proof:

We prove the statement by induction on r. For $r = 0$, it is trivial: every walk of length 0 has the same first and last vertex, and there is precisely one for each vertex. Thus

$$w_{ij}^{(0)} = \begin{cases} 1 & i = j, \\ 0 & i \neq j, \end{cases}$$

which agrees with the entries of the identity matrix $I = A(G)^0$. For the induction step, assume that the assertion holds for walks of length $r - 1$. Note that every walk W of length r consists of a walk W' of length $r - 1$ and a vertex that is adjacent to the last vertex of W'. Thus we have

$$w_{ij}^{(r)} = \sum_{k: v_k \in N(v_j)} w_{ik}^{(r-1)} = \sum_k w_{ik}^{(r-1)} a_{kj},$$

where a_{kj} is the entry in the k-th row, j-th column of $A(G)$. By the induction hypothesis, $w_{ik}^{(r-1)}$ is exactly the entry in the i-th row, k-th column of $A(G)^{r-1}$. Thus the desired statement now follows from the definition of matrix multiplication. □

5.2   Graph spectra and characteristic polynomials

The adjacency matrix, Laplacian matrix and signless Laplacian matrix are all square matrices, so it makes sense to consider their eigenvalues. The study of the spectra of these and other matrices associated with a graph is a rich area of graph theory, with many interesting applications not only in chemistry. There are also several books focusing on this subject, such as [10, 19, 99].

As a first observation, note that the spectrum of the adjacency matrix $A(G)$ is independent of the order of vertices imposed in the definition of $A(G)$. If $A'(G)$ is a second adjacency matrix of G that differs only in the order of vertices, then there exists a permutation matrix P such that

$$A'(G) = P^{-1}A(G)P.$$

Thus $A(G)$ and $A'(G)$ are similar matrices, which in particular implies that they have the same spectrum. Thus it makes sense to speak of the spectrum of a graph (i.e., the spectrum of its adjacency matrix). Likewise, the spectrum of the Laplacian matrix $L(G)$ (also called the Laplacian spectrum of G) does not depend on the order of vertices, and the same applies to the spectrum of $S(G)$ (the signless Laplacian spectrum of G).

Since $A(G)$, $L(G)$ and $S(G)$ are all symmetric, we can also immediately state the following property:

Proposition 5.2.1 *The spectrum of the adjacency matrix $A(G)$, the Laplacian matrix $L(G)$ and the signless Laplacian $S(G)$ consist entirely of real numbers. Moreover, for each of them there exists an orthogonal basis of eigenvectors with real entries.*

In view of this proposition, all eigenvalues and eigenvectors considered in the following will be real.

For example, the graph in Figure 5.1 has the adjacency matrix

$$A(G) = \begin{bmatrix} 0 & 1 & 0 & 1 \\ 1 & 0 & 1 & 1 \\ 0 & 1 & 0 & 1 \\ 1 & 1 & 1 & 0 \end{bmatrix}$$

with eigenvalues $\alpha_1 = \frac{1}{2}(1 + \sqrt{17}), \alpha_2 = \frac{1}{2}(1 - \sqrt{17})$, $\alpha_3 = 0$ and $\alpha_4 = -1$. The Laplacian matrix is

$$L(G) = \begin{bmatrix} 2 & -1 & 0 & -1 \\ -1 & 3 & -1 & -1 \\ 0 & -1 & 2 & -1 \\ -1 & -1 & -1 & 3 \end{bmatrix}$$

and has eigenvalues $\lambda_1 = \lambda_2 = 4$, $\lambda_3 = 2$ and $\lambda_4 = 0$. Finally, the signless Laplacian

$$S(G) = \begin{bmatrix} 2 & 1 & 0 & 1 \\ 1 & 3 & 1 & 1 \\ 0 & 1 & 2 & 1 \\ 1 & 1 & 1 & 3 \end{bmatrix}$$

has eigenvalues $\sigma_1 = 3 + \sqrt{5}$, $\sigma_2 = \sigma_3 = 2$ and $\sigma_4 = 3 - \sqrt{5}$.

Let us also consider the spectra associated with important families of graphs. The simplest example is the edgeless graph E_n with n vertices. In this case, $A(E_n)$, $L(E_n)$ and $S(E_n)$ are all equal to the zero matrix, which has 0 as its only eigenvalue (with multiplicity n).

The complete graph K_n is already somewhat more complicated. Its adjacency matrix has the form

$$A(K_n) = \begin{bmatrix} 0 & 1 & 1 & \cdots & 1 \\ 1 & 0 & 1 & \cdots & 1 \\ 1 & 1 & 0 & \cdots & 1 \\ \vdots & \vdots & \vdots & \ddots & \vdots \\ 1 & 1 & 1 & \cdots & 0 \end{bmatrix}.$$

Note that $A(K_n) + I$ is the matrix whose entries are all 1s. Its rank is 1, which implies that there are $n-1$ linearly independent eigenvectors for the eigenvalue -1. Thus -1 is an eigenvalue of multiplicity $n-1$, and it remains to determine the last eigenvalue. Since the trace of a matrix is the sum of its eigenvalues, and the trace is 0 in this case, the final eigenvalue must be $n-1$ (indeed, it is easy to identify the vector whose entries are all 1s as an eigenvector).

For the Laplacian, the situation is similar: it has the general form

$$L(K_n) = \begin{bmatrix} n-1 & -1 & -1 & \cdots & -1 \\ -1 & n-1 & -1 & \cdots & -1 \\ -1 & -1 & n-1 & \cdots & -1 \\ \vdots & \vdots & \vdots & \ddots & \vdots \\ -1 & -1 & -1 & \cdots & n-1 \end{bmatrix},$$

and the same argument as before shows that n is an eigenvalue of multiplicity $n-1$, with 0 being the final eigenvalue. For the signless Laplacian, which has the form

$$S(K_n) = \begin{bmatrix} n-1 & 1 & 1 & \cdots & 1 \\ 1 & n-1 & 1 & \cdots & 1 \\ 1 & 1 & n-1 & \cdots & 1 \\ \vdots & \vdots & \vdots & \ddots & \vdots \\ 1 & 1 & 1 & \cdots & n-1 \end{bmatrix},$$

the eigenvalues are $n-2$ (with multiplicity $n-1$) and $2n-2$. Let us state this as a formal proposition:

Proposition 5.2.2 *The eigenvalues of $A(K_n)$ are -1 with multiplicity $n-1$ and $n-1$, the eigenvalues of $L(K_n)$ are n with multiplicity $n-1$ and 0, and the eigenvalues of $S(K_n)$ are $n-2$ with multiplicity $n-1$ and $2n-2$.*

The eigenvalues of the matrices associated with the complete bipartite graph $K_{a,b}$ (see Section 1.2) can be obtained in a similar way. We have the following result:

Proposition 5.2.3 *The eigenvalues of $A(K_{a,b})$ are 0 with multiplicity $n-2$ (where $n = a+b$ is the number of vertices) and $\pm\sqrt{ab}$. The eigenvalues of $L(K_{a,b})$ and $S(K_{a,b})$ are a with multiplicity $b-1$, b with multiplicity $a-1$, 0 and $a+b$.*

Proof:

We observe that there are only two different types of rows in $A(K_{a,b})$, corresponding to the two partite sets. This means that the rank of $A(K_{a,b})$ is 2, while the dimension of the nullspace is $n - 2$. So there are $n - 2$ linearly independent eigenvectors for the eigenvalue 0, and it only remains to identify the two remaining eigenvalues. Consider a vector for which the entries associated with the vertices in the first partite set (of cardinality a) are all equal to x, while the remaining entries are all equal to y. For such a vector to be an eigenvector corresponding to eigenvalue α, we need to have

$$by = \alpha x \qquad \text{and} \qquad ax = \alpha y.$$

We eliminate x and y by multiplying the two equations, which gives us $ab = \alpha^2$, i.e., $\alpha = \pm\sqrt{ab}$. Possible values for x and y that yield suitable eigenvectors are $x = \sqrt{b}$ and $y = \pm\sqrt{a}$.

For the Laplacian eigenvalues, the argument is very similar. We note first that $L(K_{a,b}) - aI$ has b identical rows, thus a must be an eigenvalue of multiplicity $b - 1$. Likewise, b is an eigenvalue of multiplicity $a - 1$. This leaves us with two missing eigenvalues. The same approach as before now gives us the following two equations for the eigenvector entries x and y and the eigenvalue λ:

$$bx - by = \lambda x \qquad \text{and} \qquad ay - ax = \lambda y.$$

One solution is $\lambda = 0$ (and $x = y$), the other $\lambda = a + b = n$ (and, e.g., $x = b$, $y = -a$). $\qquad \square$

Let us now have a look at the spectra of sparse graphs, specifically trees. We will see later that the path and the star will play a major role as in previous chapters. It is thus important to know their spectra. The star is a special case of a complete bipartite graph (it can be regarded as $K_{1,n-1}$), so the following is immediate:

Corollary 5.2.1 *The eigenvalues of $A(S_n)$ are 0 with multiplicity $n - 2$ and $\pm\sqrt{n-1}$.*

For the path and the cycle, we need a little background on Chebyshev polynomials. The Chebyshev polynomials of the first kind, denoted by $T_n(x)$, are defined by the following recursion:

$$T_0(x) = 1, \ T_1(x) = x, \ T_n(x) = 2xT_{n-1}(x) - T_{n-2}(x) \text{ for } n \geq 2,$$

while the Chebyshev polynomials of the second kind satisfy the same recursion with slightly different initial values:

$$U_0(x) = 1, \ U_1(x) = 2x, \ U_n(x) = 2xU_{n-1}(x) - U_{n-2}(x) \text{ for } n \geq 2.$$

Let us list the first few Chebyshev polynomials:

$$T_0(x) = 1, \qquad\qquad\qquad U_0(x) = 1,$$
$$T_1(x) = x, \qquad\qquad\qquad U_1(x) = 2x,$$
$$T_2(x) = 2x^2 - 1, \qquad\qquad U_2(x) = 4x^2 - 1,$$
$$T_3(x) = 4x^3 - 3x, \qquad\qquad U_3(x) = 8x^3 - 4x,$$
$$T_4(x) = 8x^4 - 8x^2 + 1, \qquad U_4(x) = 16x^4 - 12x^2 + 1.$$

The Chebyshev polynomials have many remarkable properties and play a role in various areas of mathematics. We will in particular use the following connection to trigonometric functions:

Lemma 5.2.1 *For every non-negative integer n, we have*

$$T_n(\cos t) = \cos(nt) \qquad and \qquad U_n(\cos t) = \frac{\sin((n+1)t)}{\sin t}.$$

Proof:

Both identities can be proved by induction on n. Both formulas hold for $n = 0$:

$$T_0(\cos t) = U_0(\cos t) = 1 = \cos(0t) = \frac{\sin t}{\sin t},$$

as well as for $n = 1$:

$$T_1(\cos t) = \cos t, \qquad U_1(\cos t) = 2\cos t = \frac{\sin(2t)}{\sin t}.$$

The induction step is based on elementary trigonometric identities: the induction hypothesis gives us

$$\begin{aligned}
T_{n+1}(\cos t) &= 2\cos t \cdot T_n(\cos t) - T_{n-1}(\cos t) \\
&= 2\cos t \cos(nt) - \cos((n-1)t) \\
&= 2\cos t \cos(nt) - (\cos t \cos(nt) + \sin t \sin(nt)) \\
&= \cos t \cos(nt) - \sin t \sin(nt) = \cos((n+1)t)
\end{aligned}$$

and

$$\begin{aligned}
U_{n+1}(\cos t) &= 2\cos t \cdot U_n(\cos t) - U_{n-1}(\cos t) \\
&= 2\cos t \cdot \frac{\sin((n+1)t)}{\sin t} - \frac{\sin(nt)}{\sin t} \\
&= \frac{2\cos t \sin((n+1)t) - (\cos t \sin((n+1)t) - \sin t \cos((n+1)t))}{\sin t} \\
&= \frac{\cos t \sin((n+1)t) + \sin t \cos((n+1)t)}{\sin t} = \frac{\sin((n+2)t)}{\sin t}.
\end{aligned}$$

$\square$

The following lemma relates the Chebyshev polynomials to the adjacency matrices of the path and the cycle.

Lemma 5.2.2 *For every positive integer n, the characteristic polynomial of the adjacency matrix of the path P_n is*

$$\Phi_{P_n}(x) = U_n(x/2),$$

and for every integer $n \geq 3$, the characteristic polynomial of the adjacency matrix of the cycle C_n is

$$\Phi_{C_n}(x) = 2T_n(x/2) - 2.$$

Proof:

We start with the characteristic polynomial of $A(P_n)$, which is given by the $n \times n$ determinant

$$\Phi_{P_n}(x) = \begin{vmatrix} x & -1 & 0 & \cdots & 0 & 0 & 0 \\ -1 & x & -1 & \cdots & 0 & 0 & 0 \\ 0 & -1 & x & \cdots & 0 & 0 & 0 \\ \vdots & \vdots & \vdots & \ddots & \vdots & \vdots & \vdots \\ 0 & 0 & 0 & \cdots & x & -1 & 0 \\ 0 & 0 & 0 & \cdots & -1 & x & -1 \\ 0 & 0 & 0 & \cdots & 0 & -1 & x \end{vmatrix}.$$

We apply row expansion with respect to the last row, which gives us the following:

$$x \begin{vmatrix} x & -1 & 0 & \cdots & 0 & 0 \\ -1 & x & -1 & \cdots & 0 & 0 \\ 0 & -1 & x & \cdots & 0 & 0 \\ \vdots & \vdots & \vdots & \ddots & \vdots & \vdots \\ 0 & 0 & 0 & \cdots & x & -1 \\ 0 & 0 & 0 & \cdots & -1 & x \end{vmatrix} + \begin{vmatrix} x & -1 & 0 & \cdots & 0 & 0 \\ -1 & x & -1 & \cdots & 0 & 0 \\ 0 & -1 & x & \cdots & 0 & 0 \\ \vdots & \vdots & \vdots & \ddots & \vdots & \vdots \\ 0 & 0 & 0 & \cdots & x & 0 \\ 0 & 0 & 0 & \cdots & -1 & -1 \end{vmatrix}.$$

Here both determinants are of size $(n-1) \times (n-1)$. Note in particular that the first determinant is precisely the characteristic polynomial of $A(P_{n-1})$. If we further expand the second determinant with respect to the last column, we obtain

$$\Phi_{P_n}(x) = x\Phi_{P_{n-1}}(x) - \Phi_{P_{n-2}}(x).$$

Thus $\Phi_{P_n}(x)$ satisfies the same recursion as $U_n(x/2)$, and the initial values also agree: $\Phi_{P_1}(x) = x = U_1(x/2)$ and $\Phi_{P_2}(x) = x^2 - 1 = U_2(x/2)$. The first identity follows.

The characteristic polynomial of $A(C_n)$ is given by a very similar determinant, namely

$$\Phi_{C_n}(x) = \begin{vmatrix} x & -1 & 0 & \cdots & 0 & 0 & -1 \\ -1 & x & -1 & \cdots & 0 & 0 & 0 \\ 0 & -1 & x & \cdots & 0 & 0 & 0 \\ \vdots & \vdots & \vdots & \ddots & \vdots & \vdots & \vdots \\ 0 & 0 & 0 & \cdots & x & -1 & 0 \\ 0 & 0 & 0 & \cdots & -1 & x & -1 \\ -1 & 0 & 0 & \cdots & 0 & -1 & x \end{vmatrix}.$$

We use the same approach as for the path: expansion with respect to the last row gives

$$x\begin{vmatrix} x & -1 & 0 & \cdots & 0 & 0 \\ -1 & x & -1 & \cdots & 0 & 0 \\ 0 & -1 & x & \cdots & 0 & 0 \\ \vdots & \vdots & \vdots & \ddots & \vdots & \vdots \\ 0 & 0 & 0 & \cdots & x & -1 \\ 0 & 0 & 0 & \cdots & -1 & x \end{vmatrix} + \begin{vmatrix} x & -1 & 0 & \cdots & 0 & -1 \\ -1 & x & -1 & \cdots & 0 & 0 \\ 0 & -1 & x & \cdots & 0 & 0 \\ \vdots & \vdots & \vdots & \ddots & \vdots & \vdots \\ 0 & 0 & 0 & \cdots & x & 0 \\ 0 & 0 & 0 & \cdots & -1 & -1 \end{vmatrix}$$

$$+ (-1)^n \begin{vmatrix} -1 & 0 & \cdots & 0 & 0 & -1 \\ x & -1 & \cdots & 0 & 0 & 0 \\ -1 & x & \cdots & 0 & 0 & 0 \\ \vdots & \vdots & \ddots & \vdots & \vdots & \vdots \\ 0 & 0 & \cdots & x & -1 & 0 \\ 0 & 0 & \cdots & -1 & x & -1 \end{vmatrix}.$$

The first determinant is the characteristic polynomial of $A(P_{n-1})$. The second and third determinant are expanded again with respect to the last column. In each case, one of the resulting determinants is precisely the characteristic polynomial of $A(P_{n-2})$ again, while the other is the determinant of a triangular matrix that is easily determined. We find that

$$\Phi_{C_n}(x) = x\Phi_{P_{n-1}}(x) - 2\Phi_{P_{n-2}}(x) - 2.$$

The polynomials $x\Phi_{P_{n-1}}(x) - 2\Phi_{P_{n-2}}(x)$ satisfy the same recursion as $2T_n(x/2)$ (since both $\Phi_{P_{n-1}}(x)$ and $\Phi_{P_{n-2}}(x)$ do; this follows from our discussion of the path P_n above). Moreover, the initial values also agree:

$$x\Phi_{P_2}(x) - 2\Phi_{P_1}(x) = x(x^2 - 1) - 2x = x^3 - 3x = 2T_3(x/2)$$

and

$$x\Phi_{P_3}(x) - 2\Phi_{P_2}(x) = x(x^3 - 2x) - 2(x^2 - 1) = x^4 - 4x^2 + 2 = 2T_4(x/2).$$

Thus the second identity follows as well. $\square$

Now that we know the characteristic polynomials of P_n and C_n, their respective spectra are easily determined:

Proposition 5.2.4 *For every positive integer n, the spectrum of the path P_n consists of the values $2 \cos \frac{k\pi}{n+1}$, $k \in \{1, 2, \ldots, n\}$. For every integer $n \geq 3$, the spectrum of the cycle C_n consists of the values $2 \cos \frac{2k\pi}{n}$, $k \in \{0, 1, \ldots, n-1\}$.*

Proof:

Combining the previous lemma with Lemma 5.2.1, we find that

$$\Phi_{P_n}(2 \cos t) = \frac{\sin((n+1)t)}{\sin t},$$

which is zero whenever $(n+1)t$ is multiple of π, except when t is itself a multiple of π. Thus the characteristic polynomial of P_n has zeros at all values of the form $x = 2 \cos t$ corresponding to $t = \frac{k\pi}{n+1}$, where $k \in \{1, 2, \ldots, n\}$. Similarly,

$$\Phi_{C_n}(2 \cos t) = 2 \cos(nt) - 2.$$

This is zero when nt is a multiple of 2π, so we find that the zeros are $2 \cos \frac{2k\pi}{n}$, $k \in \{0, 1, \ldots, n-1\}$. $\square$

It is worth pointing out that the path P_n has n distinct eigenvalues, while the cycle has eigenvalues of multiplicity 2, since

$$2 \cos \frac{2k\pi}{n} = 2 \cos \frac{2(n-k)\pi}{n}.$$

Indeed, one can easily verify that the derivative of Φ_{C_n} at $2 \cos \frac{2k\pi}{n}$ is zero (indicating a double zero) unless $k \in \{0, n/2\}$, since the chain rule gives

$$\Phi'_{C_n}(2 \cos t) = \frac{n \sin(nt)}{\sin t}.$$

The disjoint union of graphs is perhaps the simplest possible graph operation. Later in this chapter, we will need information about the spectrum of a disjoint union of graphs. As it turns out, the spectrum of the union of two graphs is simply the union of the two spectra (taking multiplicities into account):

Proposition 5.2.5 *Let G_1 and G_2 be two (disjoint) graphs, and let H be their union. The spectrum of $A(H)$ is the union of the spectra of $A(G_1)$ and $A(G_2)$, with multiplicities added up. The same holds for the Laplacian and signless Laplacian matrices.*

Proof:

Simply note that the adjacency matrix of H has the block-diagonal shape

$$A(H) = \begin{bmatrix} A(G_1) & 0 \\ 0 & A(G_2) \end{bmatrix}$$

with respect to a suitable vertex ordering. Thus the spectrum of $A(H)$ is obtained as the union of the spectra of $A(G_1)$ and $A(G_2)$ as stated. The same also applies to the Laplacian and signless Laplacian.   □

The adjacency matrix of a graph is an important example of a non-negative matrix, so the Perron-Frobenius theorem applies to it. Let us state this theorem for later use:

Theorem 5.2.1 *Let M be a square matrix with real non-negative entries, and let ρ be its spectral radius, i.e., the greatest absolute value of an eigenvalue. The following statements hold:*

- *The spectral radius ρ is itself an eigenvalue.*

- *If $\mathbf{x}$ is a non-zero vector with non-negative entries, and the inequality*

$$M \cdot \mathbf{x} \geq \lambda \mathbf{x}$$

holds componentwise, then $\lambda \leq \rho$. If $\lambda = \rho$, then $\mathbf{x}$ must be an eigenvector, i.e., equality holds componentwise.

- *Let the entries of M be m_{ij} $(1 \leq i, j \leq n)$, and define an oriented graph with vertices $v_1, v_2, \ldots, v_n$ as follows: there is an oriented edge from v_i to v_j if and only if $m_{ij} > 0$. If this graph is strongly connected (i.e., there exists an oriented path from every vertex to every other vertex), then we call M irreducible. In this case, ρ is an eigenvalue of multiplicity 1, and there is an eigenvector associated with ρ whose entries are exclusively positive real numbers.*

The adjacency matrix $A(G)$ is an irreducible non-negative matrix if and only if G is connected. In this case, we can conclude that the spectral radius of $A(G)$ is part of the spectrum, and that there is an eigenvector with strictly positive entries associated to it. A brief introduction to the theory of the spectral radius will be provided in Section 5.8.1.

The spectra of bipartite graphs have some interesting properties that will be stated in the following two theorems. They generalize some of the patterns we observed in our examples (complete bipartite graph, star, path).

Theorem 5.2.2 *For every bipartite graph G, the spectrum of $A(G)$ is symmetric, i.e., α is an eigenvalue if and only if $-\alpha$ is. Moreover, the multiplicities of α and $-\alpha$ are the same.*

Proof:

Without loss of generality, we can assume that the vertices are ordered in such a way that the first n_1 vertices form one partite set, while the remaining n_2 vertices form the other partite set. Thus the adjacency matrix has the following block form:

$$A(G) = \begin{bmatrix} 0 & M \\ M^T & 0 \end{bmatrix}$$

for a matrix M that captures the edges between the two partite sets. Now let $\mathbf{x}$ be an eigenvector for the eigenvalue α, and split it accordingly:

$$\mathbf{x} = \begin{bmatrix} \mathbf{x}_1 \\ \mathbf{x}_2 \end{bmatrix}.$$

Then we must have

$$M \cdot \mathbf{x}_2 = \alpha \mathbf{x}_1 \quad \text{and} \quad M^T \cdot \mathbf{x}_1 = \alpha \mathbf{x}_2.$$

Equivalently,

$$M \cdot (-\mathbf{x}_2) = -\alpha \mathbf{x}_1 \quad \text{and} \quad M^T \cdot \mathbf{x}_1 = -\alpha(-\mathbf{x}_2),$$

which means that

$$\overline{\mathbf{x}} = \begin{bmatrix} \mathbf{x}_1 \\ -\mathbf{x}_2 \end{bmatrix}$$

is an eigenvector for the eigenvalue $-\alpha$. It is easy to conclude that α is an eigenvalue if and only if $-\alpha$ is one, and that the respective eigenspaces have the same dimension. The desired statement follows. □

It turns out that the converse of Theorem 5.2.2 holds as well: if the spectrum is symmetric, then the graph is bipartite. In fact, we can prove a stronger result:

Theorem 5.2.3 *Let G be a connected graph, and let ρ be the spectral radius of $A(G)$. If $-\rho$ is an eigenvalue of $A(G)$, then G must be bipartite.*

Proof:

Let $\mathbf{x}$ be an eigenvector associated with $-\rho$, and let $|\mathbf{x}|$ be the vector obtained by taking the absolute values of all entries of $\mathbf{x}$. By the triangle inequality, we have

$$\rho|\mathbf{x}| = |-\rho\mathbf{x}| = |A(G)\mathbf{x}| \leq A(G)|\mathbf{x}|$$

componentwise. By the second item in Theorem 5.2.1, $|\mathbf{x}|$ has to be an eigenvector for ρ, and equality needs to hold. Thus there cannot be any cancellations in the matrix product $A(G)\mathbf{x}$. Let us divide the vertex set into the set A of vertices for which the associated entry in $\mathbf{x}$ is positive, and the set B of vertices for which the associated entry is negative (this covers all vertices, since there cannot be any zero entries in view of the third part of Theorem 5.2.1). The only way that no cancellations occur in the product $A(G)\mathbf{x} = -\rho\mathbf{x}$ is that vertices in A are only adjacent to vertices in B, and vice versa. But this means that G is bipartite. $\square$

Corollary 5.2.2 *If the spectrum of a graph G is symmetric, i.e., whenever α is an eigenvalue of $A(G)$, then so is $-\alpha$, and the multiplicities coincide, then G is bipartite.*

Proof:

By induction on the number of connected components. If the graph is connected, we are done by the previous theorem. Otherwise, we know that the spectrum of G is the union of the spectra of its components. Consider all components for which the spectral radius ρ of $A(G)$ is an eigenvalue. Since $-\rho$ is also part of the spectrum of G, it must be in the spectrum of one of these components. It follows from the previous theorem that the component is bipartite. Now simply remove the component and invoke the induction hypothesis. $\square$

Theorem 5.2.4 *For every bipartite graph G, the spectra of the Laplacian $L(G)$ and the signless Laplacian $S(G)$ coincide.*

Proof:

The proof is very similar to that of Theorem 5.2.2. We can again assume without loss of generality that $L(G)$ and $S(G)$ have the following block forms:

$$L(G) = \begin{bmatrix} D_1 & -M \\ -M^T & D_2 \end{bmatrix} \quad \text{and} \quad S(G) = \begin{bmatrix} D_1 & M \\ M^T & D_2 \end{bmatrix}.$$

Here, D_1 and D_2 are diagonal matrices containing the vertex degrees, and M is as in the proof of Theorem 5.2.2.

Now let λ be an eigenvalue of $L(G)$, and let

$$\mathbf{x} = \begin{bmatrix} \mathbf{x}_1 \\ \mathbf{x}_2 \end{bmatrix}$$

be the associated eigenvector. Then we have

$$D_1 \cdot \mathbf{x}_1 - M \cdot \mathbf{x}_2 = \lambda\mathbf{x}_1,$$
$$-M^T \cdot \mathbf{x}_1 + D_2 \cdot \mathbf{x}_2 = \lambda\mathbf{x}_2.$$

This is equivalent to

$$\overline{\mathbf{x}} = \begin{bmatrix} \mathbf{x}_1 \\ -\mathbf{x}_2 \end{bmatrix}$$

being an eigenvector for the eigenvalue λ of $S(G)$. As in the proof of Theorem 5.2.2, it follows that λ is an eigenvalue of $L(G)$ if and only if it is an eigenvalue of $S(G)$, and the dimensions of the eigenspaces coincide. This completes the proof. $\square$

Again, a converse holds as well, and in fact we can prove a stronger statement:

Theorem 5.2.5 *Let G be a connected graph. If the spectral radius of $L(G)$ equals the spectral radius of $S(G)$, then G is bipartite.*

Proof:

Suppose that $L(G)$ and $S(G)$ have the same spectral radius ρ, and let $\mathbf{x}$ be an eigenvector of $L(G)$ for the eigenvalue ρ. Note that $S(G)$ has the same entries as $L(G)$, except that all signs are positive. Taking $|\mathbf{x}|$ to be the vector whose entries are the absolute values of the entries of $\mathbf{x}$ (as in the proof of Theorem 5.2.3), the triangle inequality gives us

$$\rho|\mathbf{x}| = |\rho\mathbf{x}| = |L(G)\mathbf{x}| \le S(G)|\mathbf{x}|,$$

so by the second item in Theorem 5.2.1, $|\mathbf{x}|$ has to be an eigenvector of $S(G)$ for the eigenvalue ρ, and equality needs to hold. Thus there cannot be any cancellations in the matrix product $L(G)\mathbf{x}$. As in the proof of Theorem 5.2.3, we can divide the vertices of G into the set A of those vertices for which the associated entries of $\mathbf{x}$ are positive and the set B of those vertices for which the entries are negative. The only way that no cancellations occur is that vertices in A are only adjacent to vertices in B, and vice versa, so G has to be bipartite. $\square$

Using the same inductive approach that gave us Corollary 5.2.2, we also obtain the following:

Corollary 5.2.3 *If the spectra of $L(G)$ and $S(G)$ coincide, then G is bipartite.*

Regular graphs, where all vertices have the same degree, are another important class in the context of spectral graph theory. We have already discussed some special cases, specifically the complete graph and the cycle. For regular graphs, the spectra of $A(G)$, $L(G)$ and $S(G)$ are very closely related, as stated in the following theorem:

Theorem 5.2.6 *Let G be a regular graph, where all vertices have degree d. If $\alpha_1, \alpha_2, \ldots, \alpha_n$ are the eigenvalues of G, then the eigenvalues of the Laplacian are $d - \alpha_1, d - \alpha_2, \ldots, d - \alpha_n$, and the eigenvalues of the signless Laplacian are $d + \alpha_1, d + \alpha_2, \ldots, d + \alpha_n$.*

Proof:

All we need to note is that the degree matrix $D(G)$ equals d times the identity matrix I. Thus we have

$$L(G) = dI - A(G) \qquad \text{and} \qquad S(G) = dI + A(G).$$

The statement of the theorem follows at once. □

Recall from Proposition 5.1.4 that the number of walks of a given length in a graph is related to the powers of the adjacency matrix. This also provides a connection to the spectrum of the adjacency matrix that is made precise in the following theorem. By a closed walk, we mean a walk whose first and last vertex are the same.

Theorem 5.2.7 *Let $\alpha_1, \alpha_2, \ldots, \alpha_n$ be the eigenvalues of the adjacency matrix of a graph G. The number of closed walks of length r in G is equal to*

$$\mathrm{tr}(A(G)^r) = \sum_{k=1}^{n} \alpha_k^r.$$

Proof:

By definition, the number of closed walks of length r in G is equal to

$$\sum_{k=1}^{n} w_{kk}^{(r)} = \mathrm{tr}(A(G)^r).$$

To obtain the formula in terms of the eigenvalues of $A(G)$, we simply combine the well-known fact that the trace of a matrix is equal to the sum of its eigenvalues with the fact that the eigenvalues of M^r are the r-th powers of the eigenvalues of M for every matrix M. □

Corollary 5.2.4 *Let $\alpha_1, \alpha_2, \ldots, \alpha_n$ be the eigenvalues of the adjacency matrix of a graph G with n vertices and m edges. We have*

$$\sum_{i=1}^{n} \alpha_i = \mathrm{tr}(A(G)) = 0$$

and

$$\sum_{i=1}^{n} \alpha_i^2 = \mathrm{tr}(A(G)^2) = 2m.$$

Proof:

The first identity is clear from the definition of the adjacency matrix, whose diagonal entries are all 0s. For the second identity, observe that walks of length

2 have the form v, w, v, where vw is an edge. Since each edge gives rise to two walks of length 2 in this way, the second identity follows. □

By definition of the Laplacian matrix $L(G)$, all its row sums are 0: the number of (-1)s in each row is exactly the number of neighbors of the corresponding vertex, which is the vertex degree, i.e., the diagonal entry. Thus the (-1)s cancel with the diagonal entry. It follows that the vector whose entries are all 1s is an eigenvector for the eigenvalue 0. We have the following theorem:

Theorem 5.2.8 *The Laplacian matrix of a graph has only non-negative eigenvalues, and 0 is always an eigenvalue. If G is connected, then the multiplicity of 0 as an eigenvalue of $L(G)$ is 1. Generally, the multiplicity of 0 as an eigenvalue of $L(G)$ is equal to the number of connected components of G.*

Proof:

As mentioned before, it is easy to see that the vector consisting only of 1s is an eigenvector for the eigenvalue 0. Now let $\mathbf{x}$ be an eigenvector of $L(G)$ for some eigenvalue λ. Without loss of generality, we may assume that at least one of its entries is positive (otherwise, consider $-\mathbf{x}$). Let x_i (corresponding to vertex v_i) be its largest entry. By definition of $L(G)$, we must have

$$\lambda x_i = \deg(v_i)x_i - \sum_{j:\, v_j \in N(v_i)} x_j \geq \deg(v_i)x_i - \sum_{j:\, v_j \in N(v_i)} x_i = 0.$$

Since $x_i > 0$ by our choice, it follows that $\lambda \geq 0$. If $\lambda = 0$, then it follows that $x_j = x_i$ for all vertices v_j that are adjacent to v_i. If G is connected, then we know that for every vertex v_k there exists a path $v_i = v_{i_1}, v_{i_2}, \ldots, v_{i_r} = v_k$ from v_i to v_k. We must have $x_i = x_{i_1} = x_{i_2}$, and repeating the argument with v_{i_2} instead of v_i shows that $x_i = x_{i_3}$ holds as well. Continuing in this way, we find that in fact $x_k = x_i$ for all k, i.e., all entries of $\mathbf{x}$ are the same. Thus there is only one eigenvector for the eigenvalue 0 if G is connected (up to multiplication by constant factors), which means that the multiplicity of 0 as an eigenvalue cannot be greater than 1. Hence it is equal to 1.

For the final statement, we can (iteratively) apply Proposition 5.2.5: each of the connected components contributes 1 to the multiplicity of 0 as an eigenvalue, so the multiplicity must generally be equal to the number of components. □

Let us now turn our attention to the characteristic polynomial: recall that we have

$$\det(xI - M) = \prod_{j=1}^{n}(x - \mu_i),$$

where $\mu_1, \mu_2, \ldots, \mu_n$ are the eigenvalues of matrix M. This simple relation explains the importance of the characteristic polynomial for the analysis of the spectrum of a matrix. As it turns out, the coefficients of the characteristic polynomial of the adjacency matrix and the Laplacian matrix of a graph can be expressed combinatorially. To this end, we need some standard results from linear algebra, such as the famous Leibniz formula for the determinant:

Lemma 5.2.3 *If M is an $n \times n$ matrix with entries m_{ij}, then we have*

$$\det M = \sum_{\pi \in S_n} \operatorname{sgn} \pi \prod_{i=1}^{n} m_{i\pi(i)},$$

where the sum is over the elements of the group S_n of all permutations of $\{1, 2, \ldots, n\}$ and $\operatorname{sgn} \pi$ is the sign of the permutation π.

Let us now consider the characteristic polynomial of the adjacency matrix: the following theorem is due to Sachs [93].

Theorem 5.2.9 *Let a Sachs graph be defined as a graph whose only connected components are single edges and cycles. For a graph G with n vertices, we let $\mathcal{U}(G)$ denote the set of all its Sachs subgraphs, which are all (not necessarily induced) subgraphs of G that are Sachs graphs, including the empty graph without vertices or edges. Moreover, let $w(H)$ denote the number of components of such a graph H, and let $c(H)$ denote the number of components that are cycles. We have*

$$\Phi_G(x) = \det(xI - A(G)) = \sum_{H \in \mathcal{U}(G)} (-1)^{w(H)} 2^{c(H)} x^{n-|H|}. \qquad (5.1)$$

Consequently, the coefficient of x^k in $\Phi_G(x)$ is

$$\sum_{\substack{H \in \mathcal{U}(G) \\ |H| = n-k}} (-1)^{w(H)} 2^{c(H)}.$$

Proof:

We apply the formula given in Lemma 5.2.3 to the matrix $M = xI - A(G)$. Every element π of the symmetric group S_n has a unique decomposition into cycles. Let us now consider the contribution of such a cycle $i_1 i_2 \ldots i_r$ to the product $\prod_{i=1}^{n} m_{i\pi(i)}$, where the entries m_{ij} are now given by

$$m_{ij} = \begin{cases} x & i = j, \\ -1 & v_i \text{ and } v_j \text{ are adjacent}, \\ 0 & \text{otherwise.} \end{cases}$$

If $r = 1$, i.e., the cycle is a singleton, then the contribution is simply a factor x. If $r > 1$, then the product over the cycle is 0 if there are some consecutive elements i_k, i_{k+1} in the cycle such that the corresponding vertices $v_{i_k}, v_{i_{k+1}}$ are not adjacent (so that one of the factors is 0). Otherwise (i.e., if the cycle in the permutation corresponds to a cycle in the graph G or a single edge, which can be interpreted as a cycle of length 2) it is $(-1)^r$, since each of the factors is -1.

So we see that the product $\prod_{i=1}^{n} m_{i\pi(i)}$ is zero unless its non-singleton cycles correspond to the components of a Sachs subgraph. Conversely, every Sachs subgraph H corresponds to $2^{c(H)}$ different permutations in S_n, since each of the cycles (of length greater than 2) can be oriented in two different ways. Let $r_1, r_2, \ldots, r_s$ (where $s = w(H)$) be the component sizes of H, and let π be one of the permutations corresponding to H. Note first that

$$|H| = \sum_{j=1}^{s} r_j.$$

Since the sign of a permutation is 1 if the number of even-length cycles is even and -1 otherwise, we have

$$\operatorname{sgn} \pi = \prod_{j=1}^{s} (-1)^{r_j+1} = (-1)^{|H|+s} = (-1)^{|H|+w(H)}.$$

Moreover,

$$\prod_{i=1}^{n} m_{i\pi(i)} = x^{n-|H|} \prod_{j=1}^{s} (-1)^{r_j} = (-1)^{|H|}.$$

It follows that

$$\operatorname{sgn} \pi \prod_{i=1}^{n} m_{i\pi(i)} = (-1)^{w(H)} x^{n-|H|}$$

for every permutation π that corresponds to H. Formula (5.1) follows immediately. □

An important corollary concerns the special case that the graph is acyclic: in this case, we find that the characteristic polynomial of (the adjacency matrix of) the graph coincides with its matching polynomial, as defined in Section 4.7. In fact, this relation is one of the reasons why the matching polynomial is defined the way it is, with alternating signs and reversed order of coefficients compared to the matching generating polynomial.

Corollary 5.2.5 *Let G be an acyclic graph (i.e., a forest). Then we have, with $m(G, k)$ denoting the number of matchings of G consisting of k edges,*

$$\Phi_G(x) = \det(xI - A(G)) = \sum_{k\geq 0} (-1)^k m(G, k) x^{n-2k}. \qquad (5.2)$$

Proof:

For an acyclic graph, Sachs subgraphs cannot contain any cycles by definition. Thus all their components are single edges, which means that every Sachs subgraph is a matching. For a Sachs subgraph H corresponding to a matching with k edges, we have $w(H) = k$ and $c(H) = 0$ in (5.1). The statement thus follows immediately from the formula in Theorem 5.2.9. $\qquad\square$

Corollary 5.2.6 *Let G be an acyclic graph with n vertices, and let $\mu(G)$ be its matching number, i.e., the maximum cardinality of a matching. Then the multiplicity of 0 as an eigenvalue of the adjacency matrix is $n - 2\mu(G)$. In particular, 0 is not an eigenvalue if and only if G has a perfect matching.*

Proof:

Simply note that the lowest exponent occurring in (5.2) with a non-zero coefficient is $x^{n-2\mu(G)}$. $\qquad\square$

The formula in Theorem 5.2.9 also simplifies significantly if T is an arbitrary bipartite graph. In this case, there are no cycles of odd length, so each component of a Sachs subgraph must have even size. So we have the following result, which is a direct consequence of Theorem 5.2.9:

Corollary 5.2.7 *Let G be a bipartite graph with n vertices. If $k \not\equiv n \mod 2$, then the coefficient of x^k in the characteristic polynomial $\det(xI - A(G))$ is 0.*

We remark that this is consistent with the statement of Theorem 5.2.2, which states that the spectrum of $A(G)$ is symmetric with respect to 0 if G is bipartite.

Recall from Theorem 4.7.1 that the roots of the matching polynomial satisfy the interlacing property. In view of Corollary 5.2.5, it follows immediately that the eigenvalues of acyclic graphs satisfy it as well. As it turns out, the interlacing property holds for all graphs. This is a consequence of the following more general statement, known as the Cauchy interlacing theorem:

Theorem 5.2.10 *Let M be a symmetric matrix, and let N be obtained from M by removing a row and the associated column. If $\mu_1, \mu_2, \ldots, \mu_n$ are the eigenvalues of M and $\nu_1, \nu_2, \ldots, \nu_{n-1}$ are the eigenvalues of N (both arranged in non-decreasing order), then*

$$\mu_1 \leq \nu_1 \leq \mu_2 \leq \nu_2 \leq \cdots \leq \mu_{n-1} \leq \nu_{n-1} \leq \mu_n.$$

Corollary 5.2.8 *Let $\alpha_1, \alpha_2, \ldots, \alpha_n$ be the eigenvalues of a graph G, and let $\beta_1, \beta_2, \ldots, \beta_{n-1}$ be the eigenvalues of $G - v$ for some vertex v (both arranged in non-decreasing order). We have*

$$\alpha_1 \leq \beta_1 \leq \alpha_2 \leq \beta_2 \leq \cdots \leq \alpha_{n-1} \leq \beta_{n-1} \leq \alpha_n.$$

Let us now consider the Laplacian matrix. In order to describe the coefficients of its characteristic polynomial, we need the notion of rooted spanning forests: a spanning forest of a graph G is a subgraph with the same vertex set that is also a forest. A rooted spanning forest is a spanning forest where each connected component has a distinguished vertex, called the root. The following theorem, first given by Kelmans [62], relates rooted spanning forests to the Laplacian matrix:

Theorem 5.2.11 *Let $r_k(G)$ be the number of rooted spanning forests of G with exactly k components. We have*

$$\Psi_G(x) = \det(xI - L(G)) = \sum_{k=1}^{n}(-1)^{n-k}r_k(G)x^k.$$

In order to prove this result, we need two important (but well-known) ingredients from linear algebra that are stated in the following lemmas:

Lemma 5.2.4 *For every $n \times n$ matrix M, the coefficient of x^{n-k} in the characteristic polynomial $\det(xI - M)$ is equal to $(-1)^k$ times the sum of all $k \times k$ minor determinants of M. In other words, $(-1)^k$ times the sum of all determinants of $k \times k$ matrices obtained by removing a set of $n - k$ rows and the corresponding $n - k$ columns from M.*

We remark that this generalizes the well-known facts that the sum of all eigenvalues is equal to the trace (which can be interpreted as the sum of all 1×1 minor determinants) and that the product of all eigenvalues is equal to the determinant of a matrix.

The following result is known as the Cauchy-Binet formula; it generalizes the formula for the determinant of a product of two square matrices.

Lemma 5.2.5 *Let A be an $n \times m$ matrix, and B an $m \times n$ matrix. For a subset S of $\{1, 2, \ldots, m\}$, let A_S be the matrix formed by the columns of A corresponding to the set S, and let B^S be the matrix formed by the rows of B corresponding to the set S. We have*

$$\det(AB) = \sum_{\substack{S \subseteq \{1,2,\ldots,m\} \\ |S|=n}} \det(A_S)\det(B^S).$$

The final ingredient that we need is the following lemma:

Lemma 5.2.6 *Let $C(T)$ be an oriented incidence matrix of a tree T, and let M be a matrix obtained from $C(T)$ by removing one of its rows. Then $\det M = \pm 1$.*

Proof:

We prove the statement by induction on the number of vertices of T. If T has two vertices, then M is a 1×1 matrix whose only entry is ± 1, so the statement holds in this case. For the induction step, pick a leaf v of T that does not correspond to the row that has been removed (this is possible since there must be at least two leaves in a tree with more than one vertex). The row that corresponds to this leaf has only one non-zero entry, in the column that corresponds to the only edge incident with v. Now perform row expansion of the determinant with respect to this row to obtain

$$\det M = \pm \det M',$$

where M' is an oriented incidence matrix of $M - v$ with one row removed. The statement now follows immediately from the induction hypothesis, and we are done. □

Theorem 5.2.11 will be an immediate consequence of the following:

Proposition 5.2.6 *For every set U of vertices, the determinant of the matrix $L(G; U)$ obtained by removing all rows and columns corresponding to U from the Laplacian matrix $L(G)$ is equal to the number of rooted spanning forests with $|U|$ components such that each of the components has a vertex in U as a root.*

Proof:

Let $C(G)$ be an oriented incidence matrix as in Proposition 5.1.3, and let M be the matrix obtained by removing all rows corresponding to vertices in U. Let n be the number of vertices in $V(G) \setminus U$, and note that the matrix $L(G; U)$ is equal to MM^T by the same argument that gave us Proposition 5.1.3. Now Lemma 5.2.5 gives us

$$\det L(G; U) = \sum_{\substack{S \subseteq \{1,2,\ldots,m\} \\ |S|=n}} \det(M_S) \det((M^T)^S).$$

Since $(M^T)^S = (M_S)^T$, this simplifies to

$$\det L(G; U) = \sum_{\substack{S \subseteq \{1,2,\ldots,m\} \\ |S|=n}} \det(M_S)^2. \tag{5.3}$$

Now we prove that $\det(M_S) = \pm 1$ if the edges corresponding to S form a spanning forest with precisely one element of U in each component, and $\det(M_S) = 0$ otherwise. To this end, we consider the following cases:

- Assume that the edges corresponding to S contain a cycle, and let $e_1, e_2, \ldots, e_k$ be the edges of that cycle. Moreover, let $\mathbf{x}_1, \mathbf{x}_2, \ldots, \mathbf{x}_k$ be the respective columns of M_S. Fix a direction for the cycle, and set $c_i = 1$ (where $i \in \{1, 2, \ldots, k\}$) if the orientation of the cycle is equal to the orientation of the edge (tail to head), and $c_i = -1$ otherwise. We have

$$\sum_{i=1}^{k} c_i \mathbf{x}_i = \mathbf{0}, \tag{5.4}$$

since the entries cancel in each row: if the row corresponds to a vertex v that is not part of the cycle formed by $e_1, e_2, \ldots, e_k$, then all entries are zero. For a vertex v that is part of the cycle, there are precisely two indices i for which the relevant entry in $c_i \mathbf{x}_i$ is non-zero (those corresponding to the two cycle edges that are incident to v), and by the choice of the coefficients c_i one of them is 1 while the other is -1. Equation (5.4) follows, which means that there is a non-trivial linear combination of columns that yields the zero vector. Thus $\det(M_S) = 0$ in this case, which is what we wanted to prove.

- If the first case does not apply, then the edges corresponding to S form an acyclic graph, i.e., a forest. Now suppose that one of the components of this forest contains two distinct elements of U, say v and w. There must be a path between v and w in this component, formed by some edges $e_1, e_2, \ldots, e_k$. For $i \in \{1, 2, \ldots, k\}$, set $c_i = 1$ if the orientation of the edge e_i is from v to w (i.e., the head of e_i is closer to w than the tail), and $c_i = -1$ otherwise. By the same argument as in the previous case, we have the identity (5.4): the entries in a row are either all equal to 0 (for vertices that are not part of the path) or cancel (for vertices on the path other than v and w). The rows corresponding to v and w would be exceptions, but they have been removed in the matrix M_S. Hence we have $\det(M_S) = 0$ again in this case.

- It remains to consider the possibility that the edges corresponding to S form a forest, and each component contains at most one element of U. If r is the number of components of the forest, then $r = |V(G)| - |S| = |V(G)| - n = |U|$, thus there is exactly one element of U in each component. Every component is a tree, so its number of vertices is equal to the number of edges plus 1. However, if we only consider vertices in $V(G) \setminus U$, then each component contains equally many vertices and edges (as there is exactly one element of U in each component). Thus we can rearrange the rows and columns of M_S in such a way that it has block diagonal form:

$$M_S = \begin{bmatrix} M_1 & 0 & \cdots & 0 \\ 0 & M_2 & \cdots & 0 \\ \vdots & \vdots & \ddots & \vdots \\ 0 & 0 & \cdots & M_r \end{bmatrix},$$

where each block M_i is a square matrix corresponding to one of the components, the rows correspond to vertices of the component (excluding the one vertex that lies in U) and the columns to edges. Some of the blocks might be empty (if the respective components are just isolated vertices), in which case we can just ignore them. Note that each of the M_i is an oriented incidence matrix of a tree with one row removed. Now we have

$$\det(M_S) = \prod_{i=1}^{r} \det(M_i),$$

and by Lemma 5.2.6 we have $\det(M_i) = \pm 1$ for all i. It follows that $\det(M_S) = \pm 1$ as well.

We see now that all terms in (5.4) that correspond to spanning forests with one element of U in each component are equal to $(\pm 1)^2 = 1$, while the other terms are all 0. The proposition follows immediately.   □

Now we have all the ingredients needed to prove the main result on the characteristic polynomial of the Laplacian:

Proof of Theorem 5.2.11:

The formula for the coefficients of the characteristic polynomial of $L(G)$ follows immediately by combining Lemma 5.2.4 with Proposition 5.2.6, summed over all sets U.   □

The special case of Proposition 5.2.6 where U consists of a single vertex gives us the following famous result, which is known as the matrix-tree theorem:

Theorem 5.2.12 *The determinant of a matrix obtained by removing a row and the corresponding column from $L(G)$ is equal to the number of spanning trees of G.*

Proof:

Simply note that a spanning forest with only one component is a spanning tree.   □

The special case $k = 1$ in Theorem 5.2.11 also yields the following formula for the total number of spanning trees:

Corollary 5.2.9 *Let G be a connected graph with n vertices, and let $\lambda_1, \lambda_2, \ldots, \lambda_n$ be its Laplacian eigenvalues, ordered in such a way that $\lambda_n = 0$ (while all other eigenvalues are positive). The number of spanning trees $t(G)$ of G is given by*

$$\frac{1}{n} \prod_{i=1}^{n-1} \lambda_i.$$

Proof:

The characteristic polynomial of the Laplacian can be factorized as

$$\Psi_G(x) = \prod_{i=1}^{n}(x - \lambda_i) = x \prod_{i=1}^{n-1}(x - \lambda_i),$$

so the coefficient of x in $\Psi_G(x)$ is $(-1)^{n-1} \prod_{i=1}^{n-1} \lambda_i$, and by Theorem 5.2.11 this is equal to $(-1)^{n-1} r_1(G)$. This means that $\prod_{i=1}^{n-1} \lambda_i$ is equal to $r_1(G)$, the number of rooted spanning forests with one component, i.e. spanning trees with a root. Since every spanning tree can be turned into a rooted spanning tree by picking any of the n vertices as a root, we have $r_1(G) = nt(G)$. The statement of the corollary follows immediately. $\square$

For example, we know from Proposition 5.2.2 that the Laplacian eigenvalues of the complete graph K_n are n with multiplicity $n-1$ and 0. Hence Corollary 5.2.9 gives us Cayley's formula: the number of spanning trees of K_n is n^{n-2}. Likewise, we find from Proposition 5.2.3 that the complete bipartite graph $K_{a,b}$ has $a^{b-1}b^{a-1}$ spanning trees.

As another remarkable corollary of Theorem 5.2.11, we obtain a connection between the Wiener index that was discussed at length in Chapter 2 and the eigenvalues of the Laplacian:

Theorem 5.2.13 *Let T be a tree with n vertices, and let $\lambda_1, \lambda_2, \ldots, \lambda_n$ be its Laplacian eigenvalues, ordered in such a way that $\lambda_n = 0$. The Wiener index of T can be expressed as*

$$W(T) = \frac{1}{n} \sum_{i=1}^{n-1} \frac{1}{\lambda_i}.$$

Proof:

Note that T only has one spanning tree, since it is a tree itself. Thus by Corollary 5.2.9, we have

$$\prod_{i=1}^{n-1} \lambda_i = n.$$

So by Theorem 5.2.11, we have

$$\prod_{i=1}^{n-1}\left(1-\frac{x}{\lambda_i}\right) = \frac{1}{\prod_{i=1}^{n-1}\lambda_i}\prod_{i=1}^{n-1}(\lambda_i - x)$$

$$= \frac{n}{x}(-1)^{n-1}\prod_{i=1}^{n}(x-\lambda_i) = \frac{n}{x}(-1)^{n-1}\Psi_G(x)$$

$$= \frac{n}{x}(-1)^{n-1}\sum_{k=1}^{n}(-1)^{n-k}r_k(T)x^k$$

$$= n\sum_{k=1}^{n}(-1)^{k-1}r_k(T)x^{k-1}.$$

Let us compare the coefficients of x on both sides of the equation: on the right side, it is simply $-nr_2(T)$. On the left side, it is

$$-\sum_{i=1}^{n-1}\frac{1}{\lambda_i}.$$

Comparing the two, we see that

$$\frac{1}{n}\sum_{i=1}^{n-1}\frac{1}{\lambda_i} = r_2(T).$$

It only remains to show that $r_2(T)$ equals the Wiener index of T. To this end, note that every spanning forest of T with two components is obtained by removing an edge e from T. For every edge $e = uv$, the number of ways of turning $T - e$ into a rooted spanning forest is equal to $n_{uv}(v) \cdot n_{uv}(u)$, where $n_{uv}(v)$ and $n_{uv}(u)$ denote the number of vertices in T closer to v and closer to u, respectively, cf. Proposition 2.2.1. This is because there are $n_{uv}(v)$ ways to pick the root of the component containing v and $n_{uv}(u)$ ways to pick the root of the component containing u. So by Proposition 2.2.1, we have

$$W(T) = \sum_{uv\in E(T)} n_{uv}(v) \cdot n_{uv}(u) = r_2(T) = \frac{1}{n}\sum_{i=1}^{n-1}\frac{1}{\lambda_i}.$$

$\square$

Theorem 5.2.13 can be found in various sources from the early 1990s. The two sides of the equation are not equal if T is an arbitrary connected graph rather than a tree. In this case, the right side of the equation, which was dubbed the quasi-Wiener index [39, 82], is equal to what is nowadays known as the Kirchhoff index [65] (the two were shown to be equal in [43]). This index is defined in an analogous way to the Wiener index by

$$\mathrm{Kf}(G) = \sum_{\{v,w\}\subseteq V(G)} r_G(v,w),$$

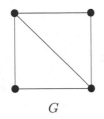
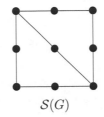

$$G \qquad\qquad\qquad S(G)$$

FIGURE 5.2
A graph and its subdivision graph.

where $r_G(v, w)$ denotes the resistance distance: it is the effective resistance between v and w in an electrical network on the graph G where each edge represents a resistor of 1 ohm. It can be shown that $r_G(v, w) \leq d_G(v, w)$ for all vertices v and w, so generally we have

$$\mathrm{Kf}(G) = \frac{1}{n} \sum_{i=1}^{n-1} \frac{1}{\lambda_i} = \sum_{\{v,w\} \subseteq V(G)} r_G(v, w) \leq \sum_{\{v,w\} \subseteq V(G)} d_G(v, w) = W(G).$$

For the signless Laplacian, the coefficients do not have an equally simple combinatorial interpretation. However, the following elegant theorem, due to Zhou and Gutman [132], relates the characteristic polynomial of the signless Laplacian to the characteristic polynomial of the adjacency matrix of the so-called subdivision graph. For a given graph G, we let the subdivision graph $S(G)$ be the graph obtained by adding a vertex on each edge, thus subdividing the edge into two edges, see Figure 5.2.

Theorem 5.2.14 *Let G be a graph with n vertices and m edges, and let $Y_G(x) = \det(xI - S(G))$ be the characteristic polynomial of its signless Laplacian. Moreover, let $\Phi_{S(G)}(x) = \det(xI - A(S(G)))$ be the characteristic polynomial of the adjacency matrix of the subdivision graph $S(G)$. The two polynomials are connected by the relation*

$$\Phi_{S(G)}(x) = x^{m-n} Y_G(x^2).$$

Proof:

The subdivision graph $S(G)$ is a bipartite graph, where the two partite sets are the old vertices that already belong to G and the new vertices correspond to the edges of G. An old and a new vertex are adjacent in $S(G)$ if and only if the corresponding vertex and edge in G are incident. Pick an order for vertices and edges of G, and let $B(G)$ be the associated incidence matrix. If we order the vertices of $S(G)$ in such a way that the old vertices come first, followed by the new vertices, while keeping the same relative order as in $B(G)$, then we obtain an adjacency matrix of the form

$$A(\mathcal{S}(G)) = \begin{bmatrix} 0 & B(G) \\ B(G)^T & 0 \end{bmatrix}.$$

Now we evaluate the determinant $\det(xI - A(\mathcal{S}(G)))$ using a special case of a technique known as Schur complement: multiplying the matrix $xI - A(\mathcal{S}(G))$ from the left by

$$\begin{bmatrix} I & x^{-1}B(G) \\ 0 & I \end{bmatrix},$$

a matrix whose determinant is 1, we obtain

$$\begin{bmatrix} I & x^{-1}B(G) \\ 0 & I \end{bmatrix} \cdot \begin{bmatrix} xI & -B(G) \\ -B(G)^T & xI \end{bmatrix} = \begin{bmatrix} xI - x^{-1}B(G)B(G)^T & 0 \\ -B(G)^T & xI \end{bmatrix}.$$

Thus

$$\begin{aligned} \det\left(xI - A(\mathcal{S}(G))\right) &= \det \begin{bmatrix} xI - x^{-1}B(G)B(G)^T & 0 \\ -B(G)^T & xI \end{bmatrix} \\ &= \det\left(xI - x^{-1}B(G)B(G)^T\right) \cdot x^m \\ &= x^{m-n} \det\left(x^2 I - B(G)B(G)^T\right). \end{aligned}$$

Now it only remains to recall that $B(G)B(G)^T = S(G)$ by Proposition 5.1.2, so the second factor is $\det(x^2 I - S(G))$, which is exactly $Y_G(x^2)$. This completes the proof. $\qquad\square$

5.3   The graph energy: elementary properties

The graph energy is one of the most important graph invariants in chemical graph theory. It was originally inspired (see [33] for the first reference) by the Hückel molecular orbital approximation, where it relates to the π-electron energy E_π. If the eigenvalues $\alpha_1, \alpha_2, \ldots, \alpha_n$ of a graph G are in non-increasing order, this quantity is given by

$$E_\pi(G) = \begin{cases} 2\sum_{i=1}^{n/2} \alpha_i & \text{if } n \text{ is even,} \\ 2\sum_{i=1}^{(n-1)/2} \alpha_i + \alpha_{(n+1)/2} & \text{if } n \text{ is odd.} \end{cases}$$

If G is bipartite, then the spectrum is symmetric by Theorem 5.2.2, so one can express this also as

$$E_\pi(G) = \sum_{k=1}^{n} |\alpha_k|.$$

This identity also holds for many non-bipartite graphs, and the difference is generally insignificant for graphs of chemical interest, which is why the right side of the equation became known as the graph energy: the energy $\text{En}(G)$ is defined as the sum of the absolute values of all eigenvalues of (the adjacency matrix of) a graph:

$$\text{En}(G) = \sum_{k=1}^{n} |\alpha_k|.$$

The notation $E(G)$ is commonly used as well, but we will rather denote the energy by $\text{En}(G)$ to distinguish it from the edge set. From a mathematical point of view, $\text{En}(G)$ is easier to handle than $E_\pi(G)$, and so it quickly became a topic of purely mathematical interest. A lot of research on the graph energy and its variants (some of which will be discussed in later sections) has been done over the decades. There is also a book devoted exclusively to the graph energy [68], and a recent monograph on its variants [38].

Let us start with an example to illustrate the concept. Consider the graph in Figure 5.1 again, which has adjacency matrix

$$A(G) = \begin{bmatrix} 0 & 1 & 0 & 1 \\ 1 & 0 & 1 & 1 \\ 0 & 1 & 0 & 1 \\ 1 & 1 & 1 & 0 \end{bmatrix}$$

and eigenvalues $\alpha_1 = \frac{1}{2}(1 + \sqrt{17}), \alpha_2 = \frac{1}{2}(1 - \sqrt{17})$, $\alpha_3 = 0$ and $\alpha_4 = -1$. Hence the energy of this graph is

$$\text{En}(G) = \left|\frac{1}{2}(1 + \sqrt{17})\right| + \left|\frac{1}{2}(1 - \sqrt{17})\right| + |0| + |-1|$$

$$= \frac{1}{2}(1 + \sqrt{17}) + \frac{1}{2}(-1 + \sqrt{17}) + 0 + 1 = \sqrt{17} + 1.$$

The formulas for the spectra of special graphs determined in Section 5.2 (specifically, Proposition 5.2.2 and Proposition 5.2.3) can be used to derive the following (simply by plugging into the definition):

Proposition 5.3.1 *The energy of the complete graph K_n is $2n - 2$, and the energy of the complete bipartite graph $K_{a,b}$ is $2\sqrt{ab}$ (in particular, the energy of the star S_n is $2\sqrt{n-1}$).*

For the path and the cycle, we also know the spectra explicitly, and as it turns out, the resulting sums simplify greatly.

Proposition 5.3.2 *The energy of the n-vertex path is*

$$\text{En}(P_n) = \begin{cases} -2 + 2\csc\left(\frac{\pi}{2n+2}\right) & n \text{ even,} \\ -2 + 2\cot\left(\frac{\pi}{2n+2}\right) & n \text{ odd,} \end{cases}$$

and the energy of the n-vertex cycle is

$$\mathrm{En}(C_n) = \begin{cases} 4\cot\frac{\pi}{n} & n \equiv 0 \mod 4, \\ 4\csc\frac{\pi}{n} & n \equiv 2 \mod 4, \\ 2\csc\frac{\pi}{2n} & n \text{ odd.} \end{cases}$$

Proof:

Recall from Proposition 5.2.4 that the eigenvalues of the path P_n are $2\cos\frac{k\pi}{n+1}$, $k \in \{1, 2, \ldots, n\}$. These are positive for $k < (n+1)/2$ and negative for $k > (n+1)/2$ (if $k = (n+1)/2$, we get 0). Thus the energy of the path is

$$\mathrm{En}(P_n) = \sum_{k=1}^{n} \left| 2\cos\frac{k\pi}{n+1} \right| = 4 \sum_{1 \le k < (n+1)/2} \cos\frac{k\pi}{n+1},$$

where the latter identity follows by symmetry. Now we write $\cos t$ as $(e^{it} + e^{-it})/2$ and apply the geometric series to obtain, first for even n,

$$\mathrm{En}(P_n) = 2\sum_{k=1}^{n/2} \left(e^{i\pi k/(n+1)} + e^{-i\pi k/(n+1)} \right)$$

$$= 2\left(\frac{e^{i\pi/(n+1)}\left(e^{i\pi n/(2n+2)} - 1 \right)}{e^{i\pi/(n+1)} - 1} + \frac{e^{-i\pi/(n+1)}\left(e^{-i\pi n/(2n+2)} - 1 \right)}{e^{-i\pi/(n+1)} - 1} \right)$$

$$= 2\left(\frac{e^{i\pi(n+2)/(2n+2)} - e^{i\pi/(n+1)}}{e^{i\pi/(n+1)} - 1} + \frac{1 - e^{-i\pi n/(2n+2)}}{e^{i\pi/(n+1)} - 1} \right)$$

$$= 2\left(-1 + \frac{e^{i\pi(n+2)/(2n+2)} - e^{-i\pi n/(2n+2)}}{e^{i\pi/(n+1)} - 1} \right)$$

$$= 2\left(-1 + \frac{e^{i\pi/2} - e^{-i\pi/2}}{e^{i\pi/(2n+2)} - e^{-i\pi/(2n+2)}} \right)$$

$$= 2\left(-1 + \frac{2i}{e^{i\pi/(2n+2)} - e^{-i\pi/(2n+2)}} \right)$$

$$= 2\left(-1 + \frac{1}{\sin(\pi/(2n+2))} \right),$$

and then analogously for odd n

$$\mathrm{En}(P_n) = 2\sum_{k=1}^{(n-1)/2} \left(e^{i\pi k/(n+1)} + e^{-i\pi k/(n+1)} \right)$$

$$= 2\left(-1 + \frac{\cos(\pi/(2n+2))}{\sin(\pi/(2n+2))} \right).$$

For the cycle C_n, the eigenvalues are $2\cos\frac{2k\pi}{n}$, $k \in \{0, 1, \ldots, n-1\}$. They are positive for $k < n/4$ and $k > 3n/4$, and negative for $n/4 < k < 3n/4$. A similar calculation as for the path gives us the stated formulas. $\square$

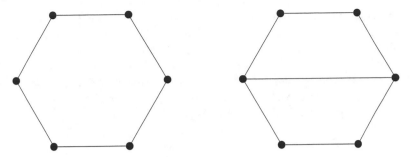

FIGURE 5.3

Counterexample to the monotonicity of the graph energy.

We remark that the following asymptotic formulas hold (obtained from the Taylor expansions of the trigonometric functions at 0):

$$\mathrm{En}(P_n) = \frac{4n}{\pi} + O(1) \qquad \text{and} \qquad \mathrm{En}(C_n) = \frac{4n}{\pi} + O(1) \qquad (5.5)$$

as $n \to \infty$.

The graph energy has the (desirable) property that it is additive over connected components, as stated in the following lemma. This is in analogy to, e.g., the fact that the Hosoya index and the Merrifield-Simmons index are multiplicative over connected components, see Lemmas 4.2.2 and 4.2.3.

Lemma 5.3.1 *Let $G_1, G_2, \ldots, G_r$ be the connected components of a graph G. The energy of G is given by*

$$\mathrm{En}(G) = \mathrm{En}(G_1) + \mathrm{En}(G_2) + \cdots + \mathrm{En}(G_r).$$

Proof:

This follows from the definition of the energy and Proposition 5.2.5. □

It is noteworthy, however, that another typical property of graph invariants is not satisfied: the analogue of Lemma 4.2.1 fails, i.e., the graph energy is not monotone with respect to insertion/deletion of edges. For example, consider the two graphs in Figure 5.3.

The six-vertex cycle, shown on the left, has spectrum $-2, -1, -1, 1, 1, 2$ (see Proposition 5.3.2) and thus an energy of 8. If we add an edge between two vertices whose distance is 3, we obtain the graph shown in the figure on the right. Its spectrum consists of ± 1 and $\pm 1 \pm \sqrt{2}$, giving us an energy of $2 + 4\sqrt{2} < 8$. Thus it is not generally true that $\mathrm{En}(G - e) < \mathrm{En}(G)$ for an edge e of G (of course, there are many instances where it is true: for example, the energy of the edgeless graph is 0, the energy of a single-edge graph is 2). This complicates in particular the search for the n-vertex graph with maximum energy. This question will be discussed later.

A remarkable property of the energy is the fact that it can be computed by means of a certain integral involving the characteristic polynomial (so-called Coulson integral formula, see [18]). There are several versions of this formula, see Theorem 5.3.1 and its corollaries below. This can not only be used to compute the energy of a given graph, but is also an extremely valuable tool in proving properties of the graph energy (and its variants, see Section 5.7). We start with a simple technical lemma.

Lemma 5.3.2 *For every real number a, we have*

$$\int_0^\infty \frac{1}{x^2} \ln(1 + a^2 x^2) \, dx = \pi |a|.$$

Proof:

The substitution $|a|x = u$ yields

$$\int_0^\infty \frac{1}{x^2} \ln(1 + a^2 x^2) \, dx = \int_0^\infty \frac{a^2}{u^2} \ln(1 + u^2) \frac{du}{|a|}$$

$$= |a| \cdot \int_0^\infty \frac{\ln(1 + u^2)}{u^2} \, du$$

$$= |a| \left(2 \arctan u - \frac{\ln(1 + u^2)}{u} \right) \Big|_0^\infty$$

$$= \pi |a|,$$

as claimed.   □

Theorem 5.3.1 *Let $\Phi_G(x) = \det(xI - A(G))$ be the characteristic polynomial of the adjacency matrix of a graph G with n vertices. The energy $\mathrm{En}(G)$ is given by the following integral:*

$$\mathrm{En}(G) = \frac{1}{\pi} \int_0^\infty \frac{1}{x^2} \ln \left(x^{2n} \Phi_G(i/x) \Phi_G(-i/x) \right) dx$$

$$= \frac{2}{\pi} \int_0^\infty \frac{1}{x^2} \left(n \ln x + \ln |\Phi_G(i/x)| \right) dx.$$

Proof:

We can write the characteristic polynomial as

$$\Phi_G(x) = \prod_{j=1}^n (x - \alpha_j),$$

where $\alpha_1, \alpha_2, \ldots, \alpha_n$ are the eigenvalues. Now note that

$$\Phi_G(i/x) \Phi_G(-i/x) = \prod_{j=1}^n (i/x - \alpha_j)(-i/x - \alpha_j) = \prod_{j=1}^n (1/x^2 + \alpha_j^2),$$

so

$$x^{2n}\Phi_G(i/x)\Phi_G(-i/x) = \prod_{j=1}^{n}(1 + \alpha_j^2 x^2)$$

and consequently

$$\int_0^\infty \frac{1}{x^2}\ln\left(x^{2n}\Phi_G(i/x)\Phi_G(-i/x)\right)dx = \sum_{j=1}^{n}\int_0^\infty \frac{1}{x^2}\ln(1 + \alpha_j^2 x^2)\,dx.$$

By Lemma 5.3.2, it follows that

$$\int_0^\infty \frac{1}{x^2}\ln\left(x^{2n}\Phi_G(i/x)\Phi_G(-i/x)\right)dx = \sum_{j=1}^{n}\pi|\alpha_j| = \pi\,\mathrm{En}(G),$$

and the first identity follows. Since $\Phi_G(x)$ is a polynomial with real coefficients, we have $\overline{\Phi_G(i/x)} = \Phi_G(\overline{i/x}) = \Phi_G(-i/x)$ for real values of x. Thus

$$\Phi_G(i/x)\Phi_G(-i/x) = |\Phi_G(i/x)|^2,$$

and we obtain

$$\mathrm{En}(G) = \frac{1}{\pi}\int_0^\infty \frac{1}{x^2}\ln\left(x^{2n}|\Phi_G(i/x)|^2\right)dx$$

$$= \frac{2}{\pi}\int_0^\infty \frac{1}{x^2}\left(n\ln x + \ln|\Phi_G(i/x)|\right)dx.$$

$\square$

The formula in Theorem 5.3.1 can be rewritten in many different ways. For instance, we have the following:

Corollary 5.3.1 *Let* $\Phi_G(x) = \det(xI - A(G))$ *be the characteristic polynomial of the adjacency matrix of a graph* G *with* n *vertices. The energy* $\mathrm{En}(G)$ *is given by the following integral:*

$$\mathrm{En}(G) = \frac{1}{\pi}\int_{-\infty}^{\infty}\left(n - \frac{ix\Phi_G'(ix)}{\Phi_G(ix)}\right)dx.$$

Proof:

We apply the substitution $1/x = u$ to the integral in Theorem 5.3.1 to obtain

$$\mathrm{En}(G) = \frac{1}{\pi}\int_0^\infty \ln\left(u^{-2n}\Phi_G(iu)\Phi_G(-iu)\right)du.$$

Integration by parts gives us

$$\mathrm{En}(G) = \frac{1}{\pi}\left(u\ln\left(u^{-2n}\Phi_G(iu)\Phi_G(-iu)\right)\Big|_0^\infty\right.$$

$$\left. - \int_0^\infty u\left(-\frac{2n}{u} + \frac{i\Phi_G'(iu)}{\Phi_G(iu)} - \frac{i\Phi_G'(-iu)}{\Phi_G(-iu)}\right)du\right). \quad (5.6)$$

Since Φ_G is a polynomial, we have $\ln\left(u^{-2n}\Phi_G(iu)\Phi_G(-iu)\right) = O(|\ln u|)$ as $u \to 0$, so the first term vanishes as $u \to 0$. Moreover, as $u \to \infty$, we know that $\Phi_G(iu) = (iu)^n + O(u^{n-2})$: the coefficient of x^{n-1} in $\Phi_G(x)$ is 0 since the sum of the eigenvalues is $\mathrm{tr}(A(G)) = 0$. Hence,

$$u^{-2n}\Phi_G(iu)\Phi_G(-iu) = 1 + O(u^{-2})$$

and consequently

$$u\ln\left(u^{-2n}\Phi_G(iu)\Phi_G(-iu)\right) = O(u^{-1}),$$

which means that the first term in (5.6) also vanishes at infinity. Thus we have

$$\begin{aligned}
\mathrm{En}(G) &= \frac{1}{\pi}\int_0^\infty \left(2n - \frac{iu\Phi_G'(iu)}{\Phi_G(iu)} + \frac{i\Phi_G'(-iu)}{\Phi_G(-iu)}\right)du \\
&= \frac{1}{\pi}\int_0^\infty \left(n - \frac{iu\Phi_G'(iu)}{\Phi_G(iu)}\right)du + \frac{1}{\pi}\int_0^\infty \left(n + \frac{iu\Phi_G'(-iu)}{\Phi_G(-iu)}\right)du.
\end{aligned}$$

We substitute $u = -v$ in the second integral and obtain

$$\begin{aligned}
\mathrm{En}(G) &= \frac{1}{\pi}\int_0^\infty \left(n - \frac{iu\Phi_G'(iu)}{\Phi_G(iu)}\right)du + \frac{1}{\pi}\int_{-\infty}^0 \left(n - \frac{iv\Phi_G'(iv)}{\Phi_G(iv)}\right)dv \\
&= \frac{1}{\pi}\int_{-\infty}^\infty \left(n - \frac{iu\Phi_G'(iu)}{\Phi_G(iu)}\right)du.
\end{aligned}$$

This completes the proof. $\qquad\square$

Another important variant of the Coulson integral formula is concerned with the comparison of two graphs.

Corollary 5.3.2 *Let $\Phi_{G_1}(x) = \det(xI - A(G_1))$ and $\Phi_{G_2}(x) = \det(xI - A(G_2))$ be the characteristic polynomials of two graphs G_1 and G_2 with n vertices. The difference of their energies can be expressed as follows:*

$$\begin{aligned}
\mathrm{En}(G_1) - \mathrm{En}(G_2) &= \frac{2}{\pi}\int_0^\infty \ln\frac{|\Phi_{G_1}(ix)|}{|\Phi_{G_2}(ix)|}\,dx \\
&= \frac{1}{\pi}\int_{-\infty}^\infty \ln\frac{|\Phi_{G_1}(ix)|}{|\Phi_{G_2}(ix)|}\,dx.
\end{aligned}$$

Proof:

We apply Theorem 5.3.1 to G_1 and G_2 and take the difference of the resulting integrals. The term $n\ln x$ cancels by the assumption that the two graphs have the same number of vertices. Thus we get

$$\mathrm{En}(G_1) - \mathrm{En}(G_2) = \frac{2}{\pi}\int_0^\infty \frac{1}{x^2}\left(\ln|\Phi_{G_1}(i/x)| - \ln|\Phi_{G_2}(i/x)|\right)dx,$$

and the first formula follows upon the substitution $x = 1/u$. The second formula is based on symmetry: $|\Phi_G(ix)| = |\Phi_G(-ix)|$ holds for every graph G and every real x, since the coefficients of the characteristic polynomial are real. $\qquad\square$

Let us now rewrite the formula of Theorem 5.3.1 one more time in terms of the coefficients of the characteristic polynomial. The result is particularly simple in the case that the graph is bipartite.

Theorem 5.3.2 *Let the characteristic polynomial of a graph G be*

$$\Phi_G(x) = \det(xI - A(G)) = \sum_{k=0}^{n} a_k x^{n-k}.$$

The energy of G can be expressed as

$$\mathrm{En}(G) = \frac{1}{\pi} \int_0^\infty \frac{1}{x^2} \ln\left(\left(\sum_{j\le n/2} (-1)^j a_{2j} x^{2j}\right)^2\right.$$
$$\left. + \left(\sum_{j\le(n-1)/2} (-1)^j a_{2j+1} x^{2j+1}\right)^2\right) dx. \quad (5.7)$$

In particular, if G is a bipartite graph, then we have

$$\mathrm{En}(G) = \frac{2}{\pi} \int_0^\infty \frac{1}{x^2} \ln\left(\sum_{j\le n/2} (-1)^j a_{2j} x^{2j}\right) dx. \quad (5.8)$$

Proof:

We note that

$$x^n \Phi_G(i/x) = \sum_{k=0}^{n} a_k i^{n-k} x^k$$
$$= i^n \left(\sum_{j\le n/2} (-1)^j a_{2j} x^{2j} - i \sum_{j\le(n-1)/2} (-1)^j a_{2j+1} x^{2j+1}\right)$$

and analogously

$$x^n \Phi_G(-i/x) = \sum_{k=0}^{n} a_k (-i)^{n-k} x^k$$
$$= i^{-n} \left(\sum_{j\le n/2} (-1)^j a_{2j} x^{2j} + i \sum_{j\le(n-1)/2} (-1)^j a_{2j+1} x^{2j+1}\right).$$

It follows that

$$x^{2n} \Phi_G(i/x)\Phi_G(-i/x) = \left(\sum_{j\le n/2} (-1)^j a_{2j} x^{2j}\right)^2 + \left(\sum_{j\le(n-1)/2} (-1)^j a_{2j+1} x^{2j+1}\right)^2,$$

so (5.7) is a direct consequence of Theorem 5.3.1. In particular, if G is bipartite, then we know from Corollary 5.2.7 that $a_{2j+1} = 0$ for all j, and (5.8) follows.   □

If G is a forest, then we can simplify even further by virtue of Corollary 5.2.5. This result will become crucial later in our analysis of trees:

Corollary 5.3.3 *If G is an acyclic graph, then we have*

$$\text{En}(G) = \frac{2}{\pi} \int_0^\infty \frac{1}{x^2} \ln \Big(\sum_{j \leq n/2} m(G,j) x^{2j} \Big) dx, \tag{5.9}$$

where $m(G,j)$ is the number of matchings of cardinality j in G.

Proof:

Simply note that $a_{2j} = (-1)^j m(G,j)$ in (5.8) by Corollary 5.2.5.   □

5.4 Bounds for the graph energy

In this section, we will prove various bounds for the energy of a graph in terms of its number of vertices and edges as well as other parameters. For our first result, we recall from Corollary 5.2.4 that we have the identities

$$\sum_{i=1}^{n} \alpha_i = \text{tr}(A(G)) = 0$$

and

$$\sum_{i=1}^{n} \alpha_i^2 = \text{tr}(A(G)^2) = 2m$$

for the eigenvalues $\alpha_1, \alpha_2, \ldots, \alpha_n$ of a graph G with n vertices and m edges.

These two identities suffice to prove a simple upper bound on the energy in terms of the number of vertices and edges that is known as the McClelland bound [76]:

Theorem 5.4.1 *For every graph G with n vertices and m edges, we have*

$$\text{En}(G) \leq \sqrt{2mn}. \tag{5.10}$$

Proof:

We can apply the inequality between the arithmetic and the quadratic mean to obtain

$$\text{En}(G)^2 = \Big(\sum_{i=1}^{n} |\alpha_i| \Big)^2 \leq n \Big(\sum_{i=1}^{n} |\alpha_i|^2 \Big) = 2mn.$$

The stated inequality follows immediately. □

We remark that equality holds in the bound (5.10) for the edgeless graph (with energy 0) and a graph whose connected components all have two vertices (connected by an edge). In the latter case, the energy is n. In fact, it can be shown that these are the only cases of equality.

It is also possible to estimate the energy of a graph only in terms of the number of edges, which is done in the following theorem:

Theorem 5.4.2 ([11]) *For every graph G with m edges, we have*

$$2\sqrt{m} \leq \text{En}(G) \leq 2m.$$

Proof:

We start with the lower bound: by some elementary manipulations, we get

$$0 = \left(\sum_{i=1}^{n} \alpha_i\right)^2 = \sum_{i=1}^{n} \alpha_i^2 + 2 \sum_{1\leq i<j\leq n} \alpha_i\alpha_j = 2m + 2 \sum_{1\leq i<j\leq n} \alpha_i\alpha_j,$$

thus

$$\sum_{1\leq i<j\leq n} \alpha_i\alpha_j = -m.$$

It follows that

$$\text{En}(G)^2 = \left(\sum_{i=1}^{n} |\alpha_i|\right)^2 = \sum_{i=1}^{n} \alpha_i^2 + 2 \sum_{1\leq i<j\leq n} |\alpha_i\alpha_j|$$

$$\geq 2m + 2\left|\sum_{1\leq i<j\leq n} \alpha_i\alpha_j\right| = 4m.$$

The lower bound $\text{En}(G) \geq 2\sqrt{m}$ follows.

For the upper bound, assume first that G has no isolated vertices. Then there can be at most $2m$ vertices (two for each edge). Combining this with the inequality in Theorem 5.4.1, we get

$$\text{En}(G) \leq \sqrt{2mn} \leq \sqrt{(2m)^2} = 2m.$$

By Theorem 5.3.1 and the fact that the energy of the graph with only one vertex is 0, adding isolated vertices does not affect the energy. Hence the bound remains true if the graph has isolated vertices. This completes the proof. □

We remark that both the upper and lower bounds are in fact sharp: for example, Proposition 5.2.3 shows that $\text{En}(G) = \sqrt{2m}$ for every complete bipartite graph $G = K_{a,b}$ with $ab = m$ (in particular, for the star with m edges), and $\text{En}(G) = 2m$ holds for every graph whose components all have either one or two vertices.

For a further refinement, we separate the greatest eigenvalue α_1 (the so-called spectral radius, which will be studied in more detail in Section 5.8.1) from the rest. In Theorem 5.8.1, we will show the following inequality for every graph with n vertices and m edges:

$$\alpha_1 \geq \frac{2m}{n}.$$

This can be used to obtain the following bound for the energy in terms of the number of vertices and the number of edges:

Theorem 5.4.3 ([66]) *For every graph G with n vertices and m edges, where $2m \geq n$, we have*

$$\mathrm{En}(G) \leq \frac{2m}{n} + \sqrt{(n-1)\left(2m - \left(\frac{2m}{n}\right)^2\right)}.$$

Proof:

Let $\alpha_1, \alpha_2, \ldots, \alpha_n$ be the eigenvalues of G, where α_1 is the greatest eigenvalue. Applying the inequality between the arithmetic and the quadratic mean in the same way as in the proof of Theorem 5.4.1, we get

$$\mathrm{En}(G) = \alpha_1 + \sum_{i=2}^{n} |\alpha_i| \leq \alpha_1 + \sqrt{(n-1)(2m - \alpha_1^2)}.$$

Now consider the derivative of the expression on the right with respect to α_1:

$$\frac{d}{dx}\left(x + \sqrt{(n-1)(2m - x^2)}\right) = 1 - \frac{x\sqrt{n-1}}{\sqrt{2m - x^2}},$$

which is positive for $0 < x < \sqrt{\frac{2m}{n}}$ and negative for $\sqrt{\frac{2m}{n}} < x < \sqrt{2m}$. Since $\alpha_1 \geq \frac{2m}{n} \geq \sqrt{\frac{2m}{n}}$ by our assumptions, the spectral radius lies in the interval for which the bound on $\mathrm{En}(G)$ is decreasing. Thus

$$\mathrm{En}(G) \leq \alpha_1 + \sqrt{(n-1)(2m - \alpha_1^2)} \leq \frac{2m}{n} + \sqrt{(n-1)\left(2m - \left(\frac{2m}{n}\right)^2\right)},$$

which completes the proof. $\square$

Now we are ready to prove the following important result on the graph energy, known as the Koolen-Moulton bound [66]:

Theorem 5.4.4 *For every graph G with n vertices, we have*

$$\mathrm{En}(G) \leq \frac{n}{2}(1 + \sqrt{n}).$$

Proof:

If $2m \leq n$, then we are done immediately by Theorem 5.4.1, since

$$\text{En}(G) \leq \sqrt{2mn} \leq n \leq \frac{n}{2}(1 + \sqrt{n}).$$

Otherwise, we can make use of the inequality in Theorem 5.4.3. Consider the bound as a function of m:

$$f(m) = \frac{2m}{n} + \sqrt{(n-1)\left(2m - \left(\frac{2m}{n}\right)^2\right)}.$$

Differentiating gives us

$$f'(m) = \frac{2}{n} - \frac{\sqrt{n-1}(4m - n^2)}{n\sqrt{2m(n^2 - 2m)}}.$$

It is not hard to verify that the derivative is positive for $m < \frac{n^2 + n^{3/2}}{4}$ and negative for $m > \frac{n^2 + n^{3/2}}{4}$. Thus we have

$$\text{En}(G) \leq f(m) \leq f\left(\frac{n^2 + n^{3/2}}{4}\right) = \frac{n}{2}(1 + \sqrt{n}).$$

$\square$

The complete graph K_4 is one of the graphs that attain the upper bound of Theorem 5.4.4. However, it is the only complete graph with this property (we already know that $\text{En}(K_n) = 2n - 2$, so this is an easy exercise). Unlike most other graph invariants (in this book and elsewhere), the complete graph is generally not extremal for the energy.

The actual cases of equality are rather more complicated: looking back over the proof, we need to have equality in all intermediate steps for equality to hold in Theorem 5.4.4. Thus $\alpha_1 = \frac{2m}{n}$, $m = \frac{n^2 + n^{3/2}}{4}$, and for equality to hold in the application of the arithmetic-quadratic mean inequality, we must have $|\alpha_2| = |\alpha_3| = \cdots = |\alpha_n|$.

The first of these conditions implies that the graph G must be regular (see Theorem 5.8.1), the second shows in particular that the number of vertices has to be a perfect square $n = k^2$, so that the number of edges is $\frac{k^4 + k^3}{4}$ and thus the degree of each vertex $\frac{k^2 + k}{2}$. The last condition is more complicated: it can be shown that it is satisfied if and only if G is a so-called strongly regular graph with parameters $(k^2, \frac{k^2 + k}{2}, \frac{k^2 + 2k}{4}, \frac{k^2 + 2k}{4})$, i.e., each of the k^2 vertices has $\frac{k^2 + k}{2}$ neighbors, any two adjacent vertices have $\frac{k^2 + 2k}{4}$ common neighbors, and any two non-adjacent vertices also have $\frac{k^2 + 2k}{4}$ common neighbors. There are infinitely many such graphs, see [66] for details.

5.5   Extremal problems in trees

Version (5.9) of the Coulson integral formula, valid for all acyclic graphs G, shows that $\mathrm{En}(G)$ is monotone in each of the coefficients $m(G, j)$, which count matchings of size j. It is therefore perhaps not surprising that the trees that are extremal with respect to the Hosoya index, which is the total number of matchings, are also typically extremal with respect to the energy. As a first instance, we have the following theorem, due to Gutman [32]:

Theorem 5.5.1 *For every tree T with n vertices, we have*

$$\mathrm{En}(S_n) \leq \mathrm{En}(T) \leq \mathrm{En}(P_n).$$

Equality holds in the first inequality if and only if T is a star, and in the second inequality if and only if T is a path.

Proof:

Recall from the discussion in Section 4.7 that S_n and P_n are extremal with respect to the partial order $\preceq_m$, defined by

$$G \preceq_m H \iff m(G, k) \leq m(H, k) \text{ for all } k,$$

in the sense that $S_n \preceq_m T$ and $T \preceq_m P_n$ for every tree T with n vertices. In both cases, there is at least one value of k such that strict inequality holds. Now simply combine this fact with the version of the Coulson integral formula given in (5.9).   □

The analogy between the Hosoya index and the energy does not only apply to the class of all trees, but also to many special families of trees. An important instance is the class of all trees with a prescribed degree sequence, as treated in Sections 2.4, 3.2 and 4.5. Here, we will not be able to work with the partial order $\preceq_m$ any longer, but the approach that we used to find the trees that maximize the Merrifield-Simmons index and minimize the Hosoya index can be used once again.

First, one observes that the proof of Lemma 4.4.6 still works when $\mu(G_{V,W}; x)$ for some fixed $x > 0$ is considered instead of $Z(G_{V,W}) = \mu(G_{V,W}, 1)$ (the definition of τ needs to be adapted accordingly). An analogue of Lemma 4.5.1 holds, and as shown by Andriantiana [3], an analogue of Theorem 4.5.4 follows as well:

Theorem 5.5.2 *For every possible degree sequence π and every fixed positive real number x, the unique tree with degree sequence π for which $\mu(T; x)$ attains its minimum is the $\mathcal{M}$-tree $\mathcal{M}(\pi)$.*

Now recall from Corollary 5.3.3 that the energy of a tree T can be expressed as

$$\mathrm{En}(T) = \frac{2}{\pi} \int_0^\infty \frac{1}{x^2} \ln\left(\mu(T; x^2)\right) dx.$$

Therefore, the following theorem follows:

Theorem 5.5.3 *For every possible degree sequence π, the unique tree with degree sequence π for which the energy attains its minimum is the $\mathcal{M}$-tree $\mathcal{M}(\pi)$.*

As in Sections 2.6.3 and 4.5, we are able to deduce a number of corollaries from Theorem 5.5.3 by means of the following majorization theorem, whose proof is analogous to that of Theorem 4.5.5:

Theorem 5.5.4 *Let π and π' be two degree sequences of trees of the same length such that π' majorizes π. If $\mathcal{M}(\pi)$ and $\mathcal{M}(\pi')$ are the $\mathcal{M}$-trees associated with π and π', then we have*

$$\mathrm{En}(\mathcal{M}(\pi')) < \mathrm{En}(\mathcal{M}(\pi)).$$

This majorization theorem yields the following corollaries, as shown in [3]:

Corollary 5.5.1 *Among all trees with n vertices and ℓ leaves, the comet obtained by attaching a path of length $n - \ell$ to the center of a star with ℓ vertices is the unique tree that minimizes the energy.*

Corollary 5.5.2 *Among all trees with n vertices and diameter D, the comet obtained by attaching a path of length $D-1$ to the center of a star with $n - D + 1$ vertices is the unique tree that minimizes the energy.*

Corollary 5.5.3 *Among all trees with n vertices and maximum degree Δ, the unique tree that attains the minimum energy is the $\mathcal{M}$-tree $\mathcal{M}(\pi_{n,\Delta})$ corresponding to the degree sequence $\pi_{n,\Delta} = (\Delta, \Delta, \ldots, \Delta, k, 1, 1, \ldots, 1)$, where the multiplicity of Δ is $\lfloor \frac{n-2}{\Delta-1} \rfloor$, and $k \in \{1, 2, \ldots, \Delta - 1\}$ is chosen to satisfy $k \equiv n - 1 \mod (\Delta - 1)$.*

5.6 Extremal problems in tree-like graphs

Once questions on the graph energy have been settled for trees, it makes sense to go one step further and consider graphs with a single cycle. A new issue that arises is the fact that the characteristic polynomial no longer coincides with the matching polynomial. However, Sachs' Theorem 5.2.9 still makes the case of unicyclic graphs manageable.

The unicyclic graph of order n whose energy is smallest coincides, perhaps unsurprisingly, with the extremal graph for the Hosoya index (at least for $n \geq 6$), cf. Theorem 4.6.2. This was shown by Hou [52].

Theorem 5.6.1 *For all integers $n \geq 6$, the unique unicyclic graph of order n with minimum energy is the graph obtained from the star S_n by adding an edge between two leaves.*

Let us remark that for $n \leq 5$, the graphs that attain the minimum are easily determined by a direct calculation: the cycles C_3 and C_4 and the graph obtained by attaching a pendant edge to a vertex of the cycle C_4.

To prove Theorem 5.6.1, we need the Coulson integral formula in the version of Theorem 5.3.2: if the characteristic polynomial of G is

$$\Phi_G(x) = \det(xI - A(G)) = \sum_{k=0}^{n} a_k x^{n-k}, \tag{5.11}$$

then the energy of G is given by

$$\text{En}(G) = \frac{1}{\pi} \int_0^\infty \frac{1}{x^2} \ln \left(\left(\sum_{j \leq n/2} (-1)^j a_{2j} x^{2j} \right)^2 \right.$$

$$\left. + \left(\sum_{j \leq (n-1)/2} (-1)^j a_{2j+1} x^{2j+1} \right)^2 \right) dx. \tag{5.12}$$

As a first step, we show that the coefficients $(-1)^j a_{2j}$ all have the same sign in unicyclic graphs, as do the coefficients $(-1)^j a_{2j+1}$.

Lemma 5.6.1 *Let G be a unicyclic graph, and let ℓ be the length of its only cycle. If ℓ is even, then $(-1)^j a_{2j} \geq 0$ for all j, and $a_{2j+1} = 0$ for all j. If $\ell \equiv 1 \mod 4$, then $(-1)^j a_{2j} \geq 0$ for all j, and $(-1)^j a_{2j+1} \leq 0$ for all j. If $\ell \equiv 3 \mod 4$, then $(-1)^j a_{2j} \geq 0$ for all j, and $(-1)^j a_{2j+1} \geq 0$ for all j.*

Proof:

If ℓ is even, then G is bipartite, which means that the spectrum is symmetric by Theorem 5.2.2. It follows that

$$\sum_{k=0}^{n} a_k (ix)^k = (ix)^n \Phi_G \left(\frac{1}{ix} \right) = \prod_{\alpha > 0} (1 + \alpha^2 x^2),$$

the product being over all positive eigenvalues. Since the coefficients of the polynomial obtained by multiplying out are clearly 0 for odd exponents and non-negative otherwise, the statement of the lemma follows in this case.

Now suppose that ℓ is odd. We apply Theorem 5.2.9, which tells us that a_k is equal to

$$\sum_{\substack{H \in \mathcal{U}(G) \\ |H|=k}} (-1)^{w(H)} 2^{c(H)},$$

the sum being over all Sachs subgraphs with k vertices ($w(H)$ is the number of components of H, and $c(H)$ the number of components that are cycles). There are two types of Sachs subgraphs in a unicyclic graph: those that contain the only cycle as a component, and those that do not. The former consist of the cycle and a matching of the remaining graph, the latter are simply matchings, as in the proof of Corollary 5.2.5. It follows immediately that

$$a_{2j} = (-1)^j m(G, j)$$

for all j, and

$$a_{2j+1} = 2(-1)^{j-(\ell-3)/2} m\left(G - C, j - \frac{\ell-1}{2}\right),$$

where C is the unique cycle of G. The statement of the lemma follows. $\quad\square$

Let us write $b_k(G)$ for the absolute value of the coefficient of x^{n-k} in the characteristic polynomial of the adjacency matrix of a graph G with n vertices, so that $b_k(G) = |a_k|$ in (5.12). By Lemma 5.6.1, we have

$$\mathrm{En}(G) = \frac{1}{\pi} \int_0^\infty \frac{1}{x^2} \ln\left(\left(\sum_{j \le n/2} b_{2j}(G) x^{2j}\right)^2\right.$$

$$\left. + \left(\sum_{j \le (n-1)/2} b_{2j+1}(G) x^{2j+1}\right)^2\right) dx. \quad (5.13)$$

Clearly, the expression on the right is monotone in all the coefficients $b_k(G)$. We consider fixed cycle length ℓ and show that the graph S_n^ℓ obtained by attaching $n - \ell$ pendant vertices to one of the vertices of the cycle C_ℓ simultaneously minimizes all these coefficients. Before we can prove this fact, we need one more auxiliary result.

Lemma 5.6.2 *Let G be a unicyclic graph, and let v be a pendant vertex of G. If u is v's unique neighbor in G, then we have*

$$b_k(G) = b_k(G - v) + b_{k-2}(G - \{u, v\})$$

for all k.

Proof:

Recall that

$$b_k(G) = \left| \sum_{\substack{H \in \mathcal{U}(G) \\ |H|=k}} (-1)^{w(H)} 2^{c(H)} \right|.$$

A Sachs subgraph H of G that contains the vertex v has to consist of the edge uv as a component and a Sachs subgraph H' of $G - \{u,v\}$. Since $w(H) = w(H') + 1$ and $c(H) = c(H')$ in this case, we find that

$$\sum_{\substack{H \in \mathcal{U}(G) \\ |H|=k}} (-1)^{w(H)} 2^{c(H)} = \sum_{\substack{H \in \mathcal{U}(G-v) \\ |H|=k}} (-1)^{w(H)} 2^{c(H)} - \sum_{\substack{H \in \mathcal{U}(G-\{u,v\}) \\ |H'|=k-2}} (-1)^{w(H')} 2^{c(H')}.$$

As shown in Lemma 5.6.1, the two sums (which are the coefficients of x^{n-k-1} and x^{n-k} in the characteristic polynomials of $G - v$ and $G - \{u,v\}$, respectively) have opposite signs (or one or both of them are equal to 0), so

$$b_k(G) = \left| \sum_{\substack{H \in \mathcal{U}(G-v) \\ |H|=k}} (-1)^{w(H)} 2^{c(H)} \right| + \left| \sum_{\substack{H \in \mathcal{U}(G-\{u,v\}) \\ |H'|=k-2}} (-1)^{w(H')} 2^{c(H')} \right|$$

$$= b_k(G - v) + b_{k-2}(G - \{u,v\}).$$

$\square$

Lemma 5.6.3 *Let G be a unicyclic graph with n vertices whose only cycle has length ℓ ($n \geq \ell \geq 3$). We have*

$$b_k(G) \geq b_k(S_n^\ell)$$

for all $k \geq 0$, and equality holds for all k only if G is isomorphic to S_n^ℓ. Consequently, $\text{En}(G) > \text{En}(S_n^\ell)$ unless G is isomorphic to S_n^ℓ.

Proof:

We prove the lemma by induction on n. For $n = \ell$, there is nothing to prove, since S_n^ℓ (isomorphic to the cycle) is the only unicyclic graph with n vertices and cycle length ℓ. For the induction step, suppose that $n > \ell$ and that G is not isomorphic to S_n^ℓ. We apply Lemma 5.6.2 to a pendant vertex v of G and its neighbor:

$$b_k(G) = b_k(G - v) + b_{k-2}(G - \{u,v\}).$$

Likewise, we have

$$b_k(S_n^\ell) = b_k(S_{n-1}^\ell) + b_{k-2}(P_{\ell-1}).$$

By the induction hypothesis, we have $b_k(S_{n-1}^\ell) \geq b_k(G - v)$. Note also that

$$b_{k-2}(P_{\ell-1}) = m\left(P_{\ell-1}, \frac{k-2}{2}\right)$$

if k is even and $k \leq \ell + 1$, and $b_{k-2}(P_{\ell-1}) = 0$ otherwise. In the latter case, the desired inequality follows immediately. If $k \leq \ell + 1$, then $k - 2 < \ell$, so a Sachs subgraph of $G - \{u, v\}$ with $k - 2$ vertices cannot contain the cycle as a component (if $G - \{u, v\}$ still contains the cycle). Thus

$$b_{k-2}(G - \{u, v\}) = m\left(G - \{u, v\}, \frac{k - 2}{2}\right).$$

Since $G - \{u, v\}$ contains $P_{\ell-1}$ as a subgraph, it follows that $b_{k-2}(G - \{u, v\}) \geq b_{k-2}(P_{\ell-1})$. Thus

$$b_k(G) \geq b_k(S_n^\ell),$$

completing the induction. Moreover, if G is not the graph S_n^ℓ, then $P_{\ell-1}$ is a proper subgraph of $G - \{u, v\}$, and we have

$$b_2(G - \{u, v\}) = |E(G - \{u, v\})| > |E(P_{\ell-1})| = b_2(P_{\ell-1}),$$

so $b_4(G) > b_4(S_n^\ell)$. The inequality $\text{En}(G) > \text{En}(S_n^\ell)$ follows from (5.13). This completes the proof. $\square$

It remains to compare the energy of S_n^ℓ for different values of ℓ.

Lemma 5.6.4 *If $n \geq \ell \geq 5$, then*

$$b_k(S_n^\ell) \geq b_k(S_n^4)$$

for all k, with strict inequality for $k = 4$. Consequently, $\text{En}(S_n^\ell) > \text{En}(S_n^4)$.

Proof:

Note first that $b_0(G) = 1$ for all graphs, and $b_2(G) = |E(G)| = n$ for all unicyclic graphs with n vertices, since the only Sachs subgraphs with two vertices are single edges. Hence the inequality holds for these two values of k. The only other value of k for which $b_k(S_n^4)$ is not zero is $k = 4$, since there are no Sachs subgraphs of S_n^4 with 1, 3 or more than 4 vertices. We easily find that

$$b_4(S_n^4) = 2n - 8,$$

since there are only two types of Sachs subgraphs: pairs of non-adjacent edges and the 4-vertex cycle. Likewise, we find that

$$b_4(S_n^\ell) = m(S_n^\ell, 2) = (\ell - 2)n - \frac{\ell^2 - \ell}{2}.$$

This is increasing in ℓ for $\ell \leq n$, so its minimum is attained for $\ell = 5$, in which case we have

$$b_4(S_n^5) = 3n - 10 > 2n - 8$$

since $n \geq 5$. The inequality is trivial whenever $k > 4$ (since $b_k(S_n^4) = 0$ in these cases), so this completes the proof. The statement on the energy follows from (5.13) again.   □

In view of Lemma 5.6.3 and Lemma 5.6.4, the only two remaining candidates for the unicyclic graph with n vertices ($n \geq 5$) that minimizes the energy are S_n^3 and S_n^4. In order to complete the proof of Theorem 5.6.1, it suffices to prove the following lemma:

Lemma 5.6.5 *For $n \geq 6$, we have* $\mathrm{En}(S_n^4) > \mathrm{En}(S_n^3)$.

Proof:

It is not difficult to determine the characteristic polynomials of S_n^3 and S_n^4, which are

$$x^n - nx^{n-2} - 2x^{n-3} + (n-3)x^{n-4}$$

and

$$x^n - nx^{n-2} + (2n-8)x^{n-4}$$

respectively. Applying the Coulson integral formula (5.13), we obtain

$$\mathrm{En}(S_n^3) = \frac{1}{\pi} \int_0^\infty \frac{1}{x^2} \ln\left(\left(1 + nx^2 + (n-3)x^4\right)^2 + \left(2x^3\right)^2\right) dx \qquad (5.14)$$

and

$$\mathrm{En}(S_n^4) = \frac{1}{\pi} \int_0^\infty \frac{1}{x^2} \ln\left(1 + nx^2 + (2n-8)x^4\right)^2 dx. \qquad (5.15)$$

Now observe that

$$\left(1 + nx^2 + (2n-8)x^4\right)^2 - \left(1 + nx^2 + (n-3)x^4\right)^2 - \left(2x^3\right)^2$$
$$= (2n-10)x^4 + (2n^2 - 10n - 4)x^6 + (3n^2 - 26n + 55)x^8.$$

For $n \geq 6$, each of the coefficients is positive, so the integrand in (5.15) is always greater than the integrand in (5.14). The stated inequality follows. □

This also concludes the proof of Theorem 5.6.1. For the maximum of the energy, the situation is somewhat more intricate, in particular the final comparison of potential candidates turns out to be rather technical. The matter was eventually resolved independently in [56] and [4], building on a number of partial results [2, 53, 55, 57]. We only state the final result.

Theorem 5.6.2 *For all $n \geq 7$ other than $9, 10, 11, 13$ and 15, the unique unicyclic graph of order n with maximum energy is the graph obtained by attaching a path of length $n - 6$ to a vertex of the 6-vertex cycle. For $n \in \{9, 10, 11, 13, 15\}$, the unique unicyclic graph of order n with maximum energy is the cycle C_n.*

We finally remark that many more results on unicyclic graphs and similar tree-like classes of graphs (for instance bicyclic graphs) can be found in the literature. We refer to the book by Li, Shi and Gutman [68] for a more detailed account.

5.7 Energy-like invariants

5.7.1 Matching energy

In this section, we study several other invariants inspired by the notion of graph energy, without going into much depth. The first of this kind is the matching energy: recall from Corollary 5.3.3 that we have

$$\text{En}(G) = \frac{2}{\pi} \int_0^\infty \frac{1}{x^2} \ln \left(\sum_{j \le n/2} m(G, j) x^{2j} \right) dx$$

if G is acyclic. However, the expression on the right is meaningful even if G is not acyclic. Generally, we call this quantity the *matching energy* and denote it by $\text{ME}(G)$. Then Corollary 5.3.3 can also be expressed as

$$\text{En}(G) = \text{ME}(G)$$

if G is acyclic. Note also that the matching energy is nothing but the sum of the absolute values of the zeros of the matching polynomial $M(G; x)$, which we know to be real by Theorem 4.7.1. To see why this holds, simply observe that we can follow the steps that gave us Corollary 5.3.3 with the characteristic polynomial replaced by the matching polynomial (of course, these coincide for acyclic graphs). To illustrate the definition, consider the graph in Figure 5.1, whose matching polynomial was earlier (in Section 4.7) determined as

$$M(G; x) = x^4 - 5x^2 + 2.$$

The roots of this polynomial are $\pm \sqrt{\frac{1}{2}(5 \pm \sqrt{17})}$, so we have

$$\text{ME}(G) = 2\sqrt{\frac{1}{2}(5 + \sqrt{17})} + 2\sqrt{\frac{1}{2}(5 - \sqrt{17})} = 2\sqrt{5 + \sqrt{8}}.$$

Indeed, it can be verified that this agrees with the integral

$$\frac{2}{\pi} \int_0^\infty \frac{1}{x^2} \ln(1 + 5x^2 + 2x^4) \, dx.$$

The notion of the matching energy was first introduced in [45], but its history actually goes further back: just like the ordinary graph energy, its roots lie in the Hückel molecular orbital approximation and the so-called π-electron energy. In some sense, the matching energy is a version of the energy that does not take cycles into account: the only Sachs subgraphs (as described in Theorem 5.2.9) that do not contain cycles are matchings, and therefore matchings represent the acyclic contribution to the characteristic polynomial. This also provides an interpretation of the fact that energy and matching energy coincide for acyclic graphs.

The difference between energy and matching energy is called the topological resonance energy, see [1, 41, 42, 45]:

$$\mathrm{TRE}(G) = \mathrm{En}(G) - \mathrm{ME}(G).$$

The matching energy shares many properties with the energy. For example, we have the following analogue of Lemma 5.3.1:

Lemma 5.7.1 *Let $G_1, G_2, \ldots, G_r$ be the connected components of a graph G. The matching energy of G is given by*

$$\mathrm{ME}(G) = \mathrm{ME}(G_1) + \mathrm{ME}(G_2) + \cdots + \mathrm{ME}(G_r).$$

Proof:

We know from Lemma 4.7.2 that

$$M(G; x) = \prod_{j=1}^{r} M(G_j; x),$$

so the zeros of $M(G; x)$ can be obtained as the union of the zeros of $M(G_1; x), M(G_2; x), \ldots, M(G_r; x)$. Thus the formula follows in the same way as Lemma 5.3.1.   □

Unlike the energy En, the matching energy ME has the useful property that it is monotone with respect to addition/subtraction of edges and vertices: to see why this is the case, note that the expression

$$\mathrm{ME}(G) = \frac{2}{\pi} \int_0^{\infty} \frac{1}{x^2} \ln \left(\sum_{j < n/2} m(G, j) x^{2j} \right) dx$$

is monotone in each of the coefficients $m(G, j)$. Clearly, none of these coefficients can increase when an edge is removed, and at least one of them (namely $m(G, 1)$, which is simply the number of edges) will decrease. The coefficients also cannot increase when a vertex is removed, and if the vertex is not isolated, then at least one edge is lost, so that $m(G, 1)$ decreases. Thus we can state the following analogue of Lemma 4.2.1:

Lemma 5.7.2 *If edges are removed from a graph, then the matching energy decreases. If vertices are removed from a graph, then the matching energy does not increase, and decreases if at least one of the vertices that are removed is not an isolated vertex.*

This also means that the question for the maximum of $\mathrm{ME}(G)$ when G is a graph with n vertices is much simpler than its counterpart for $\mathrm{En}(G)$. We can immediately state the following, in analogy, e.g., to Theorem 4.3.1 (cf. Proposition 1.7.1):

Theorem 5.7.1 *For every graph G with n vertices, we have*

$$0 = \mathrm{ME}(E_n) \leq \mathrm{ME}(G) \leq \mathrm{ME}(K_n).$$

The value of $\mathrm{ME}(K_n)$, which represents the maximum, is not given by a simple formula. It is possible, however, to provide an asymptotic formula, which we will state here without proof:

$$\mathrm{ME}(K_n) = \frac{8n^{3/2}}{3\pi} + O(n),$$

see [45] for details.

For trees, it is clear that the extremal graphs are still the path and the star, since $\mathrm{En}(T) = \mathrm{ME}(T)$ for all trees, so Theorem 5.5.1 carries over verbatim. Unicyclic and bicyclic graphs have been studied as well, see for instance [45,60].

We finish this brief discussion of the matching energy with inequalities that parallel Theorems 5.4.1 and 5.4.2.

Theorem 5.7.2 • *For every graph G with n vertices and m edges, we have* $\mathrm{ME}(G) \leq \sqrt{2mn}$.

• *For every graph G with m edges, we have* $2\sqrt{m} \leq \mathrm{ME}(G) \leq 2m$.

Proof:

Let $\mu_1, \mu_2, \ldots, \mu_n$ be the zeros of the matching polynomial of G, so that

$$\prod_{i=1}^{n}(x - \mu_i) = \sum_{k \geq 0}(-1)^k m(G, k)x^{n-2k}.$$

We compare the coefficients of x^{n-1} and x^{n-2} to obtain

$$\sum_{i=1}^{n} \mu_i = 0$$

and

$$\sum_{1 \leq i < j \leq n} \mu_i \mu_j = \frac{1}{2}\left(\sum_{i=1}^{n} \mu_i\right)^2 - \frac{1}{2}\sum_{i=1}^{n} \mu_i^2 = -m(G, 1),$$

so

$$\sum_{i=1}^{n} \mu_i^2 = 2m(G, 1) = 2m.$$

Now we apply the inequality between the arithmetic and the quadratic mean as in the proof of Theorem 5.4.1 to obtain the bound

$$\sum_{i=1}^{n} |\mu_i| \leq \sqrt{2mn}.$$

The second part follows along the same lines as Theorem 5.4.2. $\qquad\square$

5.7.2   Laplacian energy

In view of the rich theory of the graph energy, it is not unnatural to define analogous invariants corresponding to other matrices of a graph. However, the definition is not always straightforward: for the Laplacian matrix of a graph, all eigenvalues are non-negative (see Theorem 5.2.8), so taking the absolute values would not make a difference, and the sum of the Laplacian eigenvalues $\lambda_1, \lambda_2, \ldots, \lambda_n$ of a graph G is simply

$$\sum_{i=1}^{n} \lambda_i = \operatorname{tr}(L(G)) = \sum_{v \in V(G)} \deg(G) = 2m,$$

where m is the number of edges.

In view of this, the *Laplacian energy* needs to be defined in a way that differs slightly from the definition of the graph energy to become meaningful. To this end, note that the sum of the eigenvalues of the adjacency matrix is always $\operatorname{tr}(A(G)) = 0$, so the average of all eigenvalues is 0. Likewise, by the aforementioned relation, the average of all Laplacian eigenvalues is $\frac{2m}{n}$, where n is the number of vertices and m the number of edges. This motivates the following definition for the Laplacian energy [47]:

$$\operatorname{LE}(G) = \sum_{i=1}^{n} \left| \lambda_i - \frac{2m}{n} \right|.$$

Thus it measures the sum of the absolute differences from the mean, in the same way as $\operatorname{En}(G)$ is the sum of the absolute differences from the mean 0. Note that $\frac{2m}{n}$ also represents the average degree of the graph. If G is regular, then all degrees are equal to $\frac{2m}{n}$, and we can apply Theorem 5.2.6 to obtain the following:

Proposition 5.7.1 *If G is a regular graph, then* $\operatorname{LE}(G) = \operatorname{En}(G)$.

Proof:

Let d be the common degree of all vertices, and recall from Theorem 5.2.6 that the Laplacian eigenvalues are $d - \alpha_1, d - \alpha_2, \ldots, d - \alpha_n$, where $\alpha_1, \alpha_2, \ldots, \alpha_n$ are the eigenvalues of the adjacency matrix. Thus we have

$$\operatorname{LE}(G) = \sum_{i=1}^{n} \left| \lambda_i - \frac{2m}{n} \right| = \sum_{i=1}^{n} \left| \lambda_i - d \right|$$

$$= \sum_{i=1}^{n} |\alpha_i| = \operatorname{En}(G).$$

$\square$

Unlike the energy, the Laplacian energy is not generally additive in the sense of Lemma 5.3.1, the reason being that the "centralizing" term $\frac{2m}{n}$ may differ between components. For a simple example, consider the disjoint union of K_1 and K_2. Its Laplacian eigenvalues are $0, 0, 2$, so the Laplacian energy is $\frac{2}{3} + \frac{2}{3} + \frac{4}{3} = \frac{8}{3}$. On the other hand, the Laplacian energies of the two components are 0 and 2, respectively, so their sum is $2 \neq \frac{8}{3}$. However, additivity holds in a special case:

Proposition 5.7.2 *Let $G_1, G_2, \ldots, G_r$ be the connected components of a graph G, and suppose that they all have the same average degree. Then the Laplacian energy of G is*

$$\mathrm{LE}(G) = \mathrm{LE}(G_1) + \mathrm{LE}(G_2) + \cdots + \mathrm{LE}(G_r).$$

Proof:

If all components have the same average degree, then the term $\frac{2m}{n}$ in the definition of the Laplacian energy is the same for all components, so the statement follows directly from Proposition 5.2.5. $\qquad\square$

Various upper and lower bounds on the Laplacian energy, similar to those presented for the energy, have been determined in the literature. In the following, we present some basic examples.

Lemma 5.7.3 *Let $\lambda_1, \lambda_2, \ldots, \lambda_n$ be the Laplacian eigenvalues of a graph G with n vertices and m edges. We have*

$$\lambda_1^2 + \lambda_2^2 + \cdots + \lambda_n^2 = \mathrm{tr}(L(G)^2) = M_1(G) + 2m,$$

where $M_1(G)$ is the first Zagreb index, i.e., the sum of the squared degrees, see Section 3.4.

Proof:

We only need the diagonal entries of $L(G)^2$. The i-th diagonal entry is simply the inner product of the i-th row of $L(G)$ with itself, which is the sum of its squared entries. Since one of the entries is the degree $\deg(v_i)$ of the i-th vertex and there are $\deg(v_i)$ entries equal to -1 while all other entries are 0, we find that the i-th diagonal entry of $L(G)^2$ is $\deg(v_i)^2 + \deg(v_i)$. Hence

$$\mathrm{tr}(L(G)^2) = \sum_{i=1}^{n} \deg(v_i)^2 + \sum_{i=1}^{n} \deg(v_i) = M_1(G) + 2m.$$

$\square$

Corollary 5.7.1 *Let $\lambda_1, \lambda_2, \ldots, \lambda_n$ be the Laplacian eigenvalues of a graph G with n vertices and m edges. We have*

$$\sum_{i=1}^{n} \left(\lambda_i - \frac{2m}{n} \right)^2 = M_1(G) + 2m - \frac{4m^2}{n}.$$

Proof:

By Lemma 5.7.3, we have

$$\sum_{i=1}^{n} \left(\lambda_i - \frac{2m}{n}\right)^2 = \sum_{i=1}^{n} \lambda_i^2 - \frac{4m}{n} \sum_{i=1}^{n} \lambda_i + n\left(\frac{2m}{n}\right)^2$$

$$= M_1(G) + 2m - \frac{8m^2}{n} + \frac{4m^2}{n}$$

$$= M_1(G) + 2m - \frac{4m^2}{n}.$$

$\square$

The following two inequalities, due to Gutman and Zhou [47], are essentially analogues of Theorem 5.4.1 and Theorem 5.4.2 and are therefore left as exercises.

Theorem 5.7.3 *For every graph G with n vertices and m edges, we have*

$$\mathrm{LE}(G) \le \sqrt{nM_1(G) + 2mn - 4m^2}.$$

Theorem 5.7.4 *For every graph G with n vertices and m edges, we have*

$$2\sqrt{M} \le \mathrm{LE}(G) \le 2M,$$

where

$$M = M_1(G) + 2m - \frac{4m^2}{n}.$$

As pointed out by Robbiano and Jiménez [92], a general theorem due to Ky Fan [29] can be used to obtain inequalities for the Laplacian energy. Ky Fan's theorem can be stated as follows: for a square matrix M, let $\mathcal{E}(M)$ be the sum of its singular values (for symmetric matrices, this is precisely the sum of the absolute values of the eigenvalues). The following inequality holds for all square matrices A, B of the same size:

$$\mathcal{E}(A + B) \le \mathcal{E}(A) + \mathcal{E}(B). \tag{5.16}$$

Among other things, this can be used to obtain a bound on the Laplacian energy in terms of the number of vertices and the number of edges only:

Theorem 5.7.5 *For every graph G with n vertices and m edges, we have*

$$LE(G) \le 4m\left(1 - \frac{1}{n}\right).$$

Proof:

Let $\lambda_1, \lambda_2, \ldots, \lambda_n$ be the Laplacian eigenvalues of G, and note that $\lambda_1 - \frac{2m}{n}, \lambda_2 - \frac{2m}{n}, \ldots, \lambda_n - \frac{2m}{n}$ are the eigenvalues of $L(G) - \frac{2m}{n}I$. Therefore, we have

$$\text{LE}(G) = \mathcal{E}\left(L(G) - \frac{2m}{n}I\right).$$

We write the matrix $L(G) - \frac{2m}{n}I$ as a sum of smaller matrices, one for each edge of G. Suppose that the ends of an edge e are the vertices v_k and v_ℓ. We define the matrix L_e by its entries $\ell_{ij}(e)$ as follows:

$$\ell_{ij}(e) = \begin{cases} 1 & i = j = k \text{ or } i = j = \ell, \\ -1 & i = k \text{ and } j = \ell \text{ or } i = \ell \text{ and } j = k, \\ 0 & \text{otherwise.} \end{cases}$$

Note that

$$L(G) - \frac{2m}{n}I = \sum_{e \in E(G)} \left(L_e - \frac{2}{n}I\right),$$

so Ky Fan's inequality (5.16) (iterated to apply to multiple summands) yields

$$\text{LE}(G) \leq \sum_{e \in E(G)} \mathcal{E}\left(L_e - \frac{2}{n}I\right).$$

It is easy to see that the eigenvalues of L_e are 2 and 0 (with multiplicity $n-1$) for every edge e. Hence we have

$$\mathcal{E}\left(L_e - \frac{2}{n}I\right) - 2 - \frac{2}{n} + (n-1) \cdot \frac{2}{n} = 4 - \frac{4}{n}.$$

The desired inequality follows immediately. □

We conclude this section on the Laplacian energy with a proof of the fact that the maximum Laplacian energy among trees is attained by the star. While this might not seem surprising at this point, it is worth mentioning that the analogous problem for the minimum is still unsolved. As one would expect, the path is conjectured to yield the minimum. The proof is based on the following result due to Fritscher, Hoppen, Rocha and Trevisan [31], whose rather involved proof we skip.

Theorem 5.7.6 *For every tree T with n vertices, and every $k \in \{1, 2, \ldots, n\}$, the sum of the k largest Laplacian eigenvalues of T is less than or equal to*

$$n + 2k - 2 - \frac{2k - 2}{n}.$$

Equality holds if and only if $k = 1$ and T is a star.

The following is now a direct consequence:

Theorem 5.7.7 *For every tree T with n vertices, we have*

$$\text{LE}(T) \leq 2n - 4 + \frac{4}{n},$$

with equality if and only if T is a star.

Proof:

Let the Laplacian eigenvalues of T be $\lambda_1, \lambda_2, \ldots, \lambda_n$, sorted in non-increasing order. Let k be the largest index for which $\lambda_k \geq \frac{2m}{n}$ (where $m = n - 1$ is the number of edges). We have

$$
\begin{aligned}
\text{LE}(T) &= \sum_{i=1}^{n} \left| \lambda_i - \frac{2m}{n} \right| \\
&= \sum_{i=1}^{k} \left(\lambda_i - \frac{2m}{n} \right) + \sum_{i=k+1}^{n} \left(\frac{2m}{n} - \lambda_i \right) \\
&= \sum_{i=1}^{k} \lambda_i - \frac{2km}{n} + \frac{2(n-k)m}{n} - \sum_{i=1}^{n} \lambda_i + \sum_{i=1}^{k} \lambda_i \\
&= 2 \sum_{i=1}^{k} \lambda_i + \frac{2(n-2k)m}{n} - 2m \\
&= 2 \sum_{i=1}^{k} \lambda_i - \frac{4km}{n}.
\end{aligned}
$$

Now apply Theorem 5.7.6 and plug in $m = n - 1$ to obtain

$$\text{LE}(T) \leq 2 \left(n + 2k - 2 - \frac{2k-2}{n} \right) - \frac{4k(n-1)}{n} = 2n - 4 + \frac{4}{n},$$

completing the proof of the inequality. By Theorem 5.7.6, equality can only hold for the star, and it is easy to verify that the star S_n does indeed have Laplacian energy $2n - 4 + \frac{4}{n}$. □

5.7.3 Incidence energy and Laplacian-energy-like invariant

The sum of the absolute values of the eigenvalues is in principle a well-defined quantity for every square matrix. However, for non-square matrices it does not make sense. A way to generalize to arbitrary matrices is to use the singular values instead of the eigenvalues. Recall here that the singular values of a (real-valued) matrix M are the square roots of the eigenvalues of the matrix $M^T M$. Since $M^T M$ is always symmetric and positive semidefinite, its eigenvalues are non-negative, so the singular values are non-negative real numbers. Moreover, it is worth pointing out the well-known result that $M^T M$ and $M M^T$ always have the same eigenvalues, except for some additional zeros in the spectrum of the larger matrix.

Definition 5.7.1 *The energy of a (not necessarily square) matrix is the sum of its singular values.*

If M is already a symmetric matrix, then the eigenvalues of $M^T M$ are the squares of the eigenvalues of M, so the singular values are simply the absolute values of M's eigenvalues. Thus the energy of $A(G)$ as given in Definition 5.7.1 coincides with the definition of the graph energy. Having a definition of energy at our disposal for arbitrary matrices, it makes sense now to define the *incidence energy* $\text{IE}(G)$ as the energy associated with the incidence matrix $B(G)$. The first paper on this version of the energy is due to Jooyandeh, Kiani and Mirzakhah [61].

Now recall from Proposition 5.1.2 that $B(G)B(G)^T = S(G)$, so the non-zero singular values of $B(G)$ are just the non-zero eigenvalues of the signless Laplacian, and we can say the following:

Proposition 5.7.3 *The incidence energy of a graph is equal to the sum of the square roots of the eigenvalues of its signless Laplacian.*

Recall also from Proposition 5.2.4 that the spectrum of the Laplacian is equal to the spectrum of the signless Laplacian if the graph is bipartite. Hence the following is immediate:

Proposition 5.7.4 *If G is a bipartite graph, then $\text{IE}(G)$ is equal to the sum of the square roots of its Laplacian eigenvalues.*

For example, if we consider the graph in Figure 5.1, whose incidence matrix is

$$B(G) = \begin{bmatrix} 1 & 0 & 1 & 0 & 0 \\ 1 & 1 & 0 & 0 & 1 \\ 0 & 1 & 0 & 1 & 0 \\ 0 & 0 & 1 & 1 & 1 \end{bmatrix},$$

we find that the singular values are (recall the eigenvalues of the signless Laplacian that were determined earlier) $\sqrt{3+\sqrt{5}} = \frac{1+\sqrt{5}}{\sqrt{2}}$, $\sqrt{2}$ with multiplicity 2 and $\sqrt{3-\sqrt{5}} = \frac{-1+\sqrt{5}}{\sqrt{2}}$. Consequently, the incidence energy in this example is $\text{IE}(G) = \sqrt{10} + \sqrt{8}$.

The quantity that occurs in Proposition 5.7.4 is known as the *Laplacian-energy-like invariant* (LEL for short): if $\lambda_1, \lambda_2, \ldots, \lambda_n$ are the Laplacian eigenvalues of G, then we set

$$\text{LEL}(G) = \sum_{i=1}^{n} \sqrt{\lambda_i}.$$

This was first proposed by Liu and Liu [72]. In some sense, the Laplacian-energy-like invariant is mathematically better behaved than the Laplacian energy. For example, it satisfies the additivity property of Lemma 5.3.1 that the Laplacian energy does not. This is also true for the incidence energy.

Lemma 5.7.4 *Let $G_1, G_2, \ldots, G_r$ be the connected components of a graph G. The incidence energy and the Laplacian-energy-like invariant of G satisfy the identities*

$$\text{IE}(G) = \text{IE}(G_1) + \text{IE}(G_2) + \cdots + \text{IE}(G_r)$$

and

$$\text{LEL}(G) = \text{LEL}(G_1) + \text{LEL}(G_2) + \cdots + \text{LEL}(G_r).$$

Proof:

This follows from Proposition 5.2.5 in the same way that Lemma 5.3.1 was obtained.   □

At the end of Section 5.2, we proved a connection between the signless Laplacian and the adjacency matrix of the subdivision graph. We can use this connection now to prove a relation between the incidence energy and the "ordinary" energy:

Theorem 5.7.8 *For every graph G, the energy of the subdivision graph $\mathcal{S}(G)$ equals twice the incidence energy of G:*

$$\text{En}(\mathcal{S}(G)) = 2\,\text{IE}(G).$$

Proof:

We know from Theorem 5.2.14 that the characteristic polynomial of the adjacency matrix of $\mathcal{S}(G)$ and the characteristic polynomial of the signless Laplacian $\mathcal{S}(G)$ are connected by the relation

$$\det(xI - A(\mathcal{S}(G))) = \Phi_{\mathcal{S}(G)}(x) = x^{m-n} Y_G(x^2) = x^{m-n} \det(x^2 I - S(G)).$$

So if $\sigma_1, \sigma_2, \ldots, \sigma_n$ are the eigenvalues of $S(G)$, then the eigenvalues of $A(\mathcal{S}(G))$ are precisely $\pm\sqrt{\sigma_1}, \pm\sqrt{\sigma_2}, \ldots, \pm\sqrt{\sigma_n}$, except for a number of zeros. Since the zeros do not contribute to either $\text{En}(\mathcal{S}(G))$ or $\text{IE}(G)$, the identity follows.   □

For both the incidence energy and the Laplacian-energy-like invariant, we also have an integral representation of the same form as in Theorem 5.3.1 for the ordinary graph energy.

Theorem 5.7.9 *Let $\Psi_G(x) = \det(xI - L(G))$ and $Y_G(x) = \det(xI - S(G))$ be the characteristic polynomials of the Laplacian and signless Laplacian of a graph G with n vertices. The invariants $\text{LEL}(G)$ and $\text{IE}(G)$ are given by the following integrals:*

$$\text{LEL}(G) = \frac{1}{\pi} \int_0^\infty \frac{1}{x^2} \ln\left(x^{2n}|\Psi_G(-1/x^2)|\right) dx$$

$$= \frac{1}{2\pi} \int_0^\infty \frac{1}{\sqrt{u}} \ln\left(u^{-n}|\Psi_G(-u)|\right) du$$

and

$$\mathrm{IE}(G) = \frac{1}{\pi} \int_0^\infty \frac{1}{x^2} \ln\left(x^{2n}|Y_G(-1/x^2)|\right) dx$$
$$= \frac{1}{2\pi} \int_0^\infty \frac{1}{\sqrt{u}} \ln\left(u^{-n}|Y_G(-u)|\right) du.$$

Proof:

We only consider the formulas for $\mathrm{LEL}(G)$, the proof of the second part is completely analogous. Let $\lambda_1, \lambda_2, \dots, \lambda_n$ be the eigenvalues of $L(G)$, and note that the zeros of $\Psi_G(x^2)$ are precisely $\pm\sqrt{\lambda_1}, \pm\sqrt{\lambda_2}, \dots, \pm\sqrt{\lambda_n}$. So the sum of the absolute values of the zeros of $\Psi_G(x^2)$ is precisely equal to $2\,\mathrm{LEL}(G)$. Thus we obtain a formula for $\mathrm{LEL}(G)$ by replacing $\Phi_G(x)$ by $\Psi_G(x^2)$ and n by $2n$ in the formula given in Theorem 5.3.1, and finally dividing by 2. The first formula follows immediately, and the second is obtained by means of the substitution $x^{-2} = u$. $\square$

Recall from Theorem 5.2.11 that the characteristic polynomial of the Laplacian is given by

$$\Psi_G(x) = \det(xI - L(G)) = \sum_{k=1}^n (-1)^{n-k} r_k(G) x^k,$$

where $r_k(G)$ is the number of rooted spanning forests of G with exactly k components. Plugging this into the formula given in Theorem 5.7.9, we obtain

$$\mathrm{LEL}(G) = \frac{1}{2\pi} \int_0^\infty \frac{1}{\sqrt{u}} \ln\left(\sum_{k=1}^n r_k(G) u^{k-n}\right) du. \qquad (5.17)$$

The integral is clearly an increasing function in each of the coefficients $r_k(G)$. Since adding an edge cannot decrease the number of rooted spanning forests with any given number of components and also increases the total number of rooted spanning forests (there are always spanning forests including the new edge), the following monotonicity property is immediate:

Lemma 5.7.5 *If edges are added to a graph, then the LEL increases. If edges are removed from a graph, then the LEL decreases.*

The following theorem is now an immediate consequence:

Theorem 5.7.10 *For every graph G with n vertices, we have*

$$0 = \mathrm{LEL}(E_n) \le \mathrm{LEL}(G) \le \mathrm{LEL}(K_n) = (n-1)\sqrt{n}.$$

Equality in the first inequality only holds if G is edgeless, and equality in the second inequality only holds if G is complete.

Proof:

The Laplacian spectrum of the edgeless graph only consists of zeros, while the Laplacian spectrum of K_n consists of n (with multiplicity $n-1$) and 0, see Proposition 5.2.2. Thus we have $\text{LEL}(E_n) = 0$ and $\text{LEL}(K_n) = (n-1)\sqrt{n}$. The inequalities are immediate from the monotonicity property in Lemma 5.7.5, so the theorem follows.   □

For the incidence energy, we cannot argue in the same way as for the LEL, since the coefficients of the characteristic polynomial of the signless Laplacian do not have an equally simple combinatorial interpretation. Instead, we follow a different approach that also establishes a nice connection to the line graph (see for instance [37]).

Lemma 5.7.6 *Let e be an arbitrary edge of a graph G. We have*

$$\text{IE}(G - e) < \text{IE}(G).$$

Proof:

The proof is based on the fact mentioned at the beginning of this section that $M^T M$ and $M M^T$ always have the same eigenvalues, except for some additional zeros in the spectrum of the larger matrix. We apply this to the incidence matrix $B(G)$ of the graph G: since $S(G) = B(G)B(G)^T$ by Proposition 5.1.2 and $B(G)^T B(G) = A(\mathcal{L}(G)) + 2I$ by Proposition 5.1.1, the eigenvalues of the signless Laplacian $S(G)$ are, up to additional zeros, the eigenvalues of $A(\mathcal{L}(G)) + 2I$. Thus if $\alpha_1, \alpha_2, \ldots, \alpha_m$ are the eigenvalues of $A(\mathcal{L}(G))$, then the eigenvalues of $S(G)$ are $\alpha_1 + 2, \alpha_2 + 2, \ldots, \alpha_m + 2$ (up to the number of zeros), and it follows that

$$\text{IE}(G) = \sum_{i=1}^{m} \sqrt{\alpha_i + 2}.$$

Now note that $\mathcal{L}(G - e)$ is obtained from $\mathcal{L}(G)$ by removing a vertex (the vertex corresponding to e). So by the interlacing property of Corollary 5.2.8, the eigenvalues $\beta_1, \beta_2, \ldots, \beta_{m-1}$ of $A(\mathcal{L}(G - e))$ satisfy

$$\alpha_1 \leq \beta_1 \leq \alpha_2 \leq \beta_2 \leq \cdots \leq \alpha_{m-1} \leq \beta_{m-1} \leq \alpha_m.$$

It follows that

$$\text{IE}(G - e) = \sum_{i=1}^{m-1} \sqrt{\beta_i + 2} \leq \sum_{i=2}^{m} \sqrt{\alpha_i + 2} \leq \sum_{i=1}^{m} \sqrt{\alpha_i + 2} = \text{IE}(G). \quad (5.18)$$

For equality to hold, we would have to have $\alpha_1 + 2 = 0$ and $\alpha_i = \beta_{i-1}$ for all $i > 1$. However, since

$$\sum_{i=1}^{m}(\alpha_i + 2) = \operatorname{tr}(A(\mathcal{L}(G)) + 2I) = 2m$$

$$> 2(m-1) = \operatorname{tr}(A(\mathcal{L}(G-e)) + 2I) = \sum_{i=2}^{m}(\beta_{i-1} + 2),$$

this is impossible. Thus we have strict inequality in (5.18). $\qquad\square$

Again, bounds for $\mathrm{IE}(G)$ follow in the same way as Theorem 5.7.10:

Theorem 5.7.11 *For every graph G with n vertices, we have*

$$0 = \mathrm{IE}(E_n) \leq \mathrm{IE}(G) \leq \mathrm{IE}(K_n) = (n-1)\sqrt{n-2} + \sqrt{2n-2}.$$

Equality in the first inequality only holds if G is edgeless, and equality in the second inequality only holds if G is complete.

It was mentioned earlier that $\mathrm{IE}(G)$ and $\mathrm{LEL}(G)$ coincide for bipartite graphs. This is not the case for arbitrary graphs (the first simple counterexample being the complete graph K_3), but we will prove that $\mathrm{IE}(G) \geq \mathrm{LEL}(G)$ for all graphs G. To this end, we need the following result on the coefficients of the respective characteristic polynomials:

Lemma 5.7.7 *Let G be a graph with n vertices, and let the characteristic polynomials of the Laplacian and signless Laplacian be*

$$\Psi_G(x) = \det(xI - L(G)) = \sum_{k=0}^{n}(-1)^{n-k}a_k x^k$$

and

$$Y_G(x) = \det(xI - S(G)) = \sum_{k=0}^{n}(-1)^{n-k}b_k x^k,$$

respectively. For all $k \in \{0, 1, \ldots, n\}$, we have $a_k \leq b_k$.

Proof:

Recall from the proof of Theorem 5.2.11 that the coefficient a_k is equal to the sum $\sum_{|U|=k} \det L(G; U)$ of all determinants of matrices of the form $L(G; U)$, obtained by removing all rows and columns corresponding to the vertex set U from the Laplacian $L(G)$. Moreover, if we fix U, then the Cauchy-Binet formula gives us

$$\det L(G; U) = \sum_{\substack{S \subseteq \{1,2,\ldots,m\} \\ |S|=n}} \det(M_S)^2, \qquad (5.19)$$

where M is obtained from the oriented incidence matrix $C(G)$ by removing all rows corresponding to U, m is the number of edges and n the number of vertices in $V(G) \setminus U$. We can use the same reasoning to find that b_k is equal to the sum $\sum_{|U|=k} \det S(G; U)$, where $S(G; U)$ is constructed from $S(G)$ in the same way that $L(G; U)$ is constructed from $L(G)$. Moreover, we also have

$$\det S(G; U) = \sum_{\substack{S \subseteq \{1,2,\ldots,m\} \\ |S|=n}} \det(N_S)^2, \tag{5.20}$$

where N is obtained from the incidence matrix $B(G)$ in the same way that M is obtained from $C(G)$.

In the proof of Theorem 5.2.11, we showed that $\det(M_S) \in \{-1, 0, 1\}$ for all S, so that $\det(M_S)^2$ is either 0 or 1. Since N_S has only integer entries, its determinant is always an integer, and $\det(N_S)^2$ is always a non-negative integer. If we can show that $\det M_S = 0$ whenever $\det N_S = 0$, then it follows that $\det(N_S)^2 \geq \det(M_S)^2$ for all S, thus

$$\det L(G; U) \leq \det S(G; U)$$

by (5.19) and (5.20). The inequality $a_k \leq b_k$ then follows by summing over all vertex sets U.

To show that $\det N_S = 0$ implies $\det M_S = 0$, we use a parity argument: the entries of M_S only differ from the entries of N_S in some signs. Now apply the Leibniz formula (Lemma 5.2.3) to M_S and to N_S. The terms in the two resulting sums are the same except for the signs, and each of them is an integer. Therefore, the difference between $\det M_S$ and $\det N_S$ has to be an even integer. Since $\det M_S$ can only be 0 or ± 1, we have $\det M_S = 0$ whenever $\det N_S$ is an even integer, in particular whenever $\det N_S = 0$. This completes the proof. $\square$

Now the aforementioned inequality between incidence energy and LEL follows:

Theorem 5.7.12 *For every graph G, we have*

$$\mathrm{IE}(G) \geq \mathrm{LEL}(G),$$

and equality holds if and only if G is bipartite.

Proof:

Writing

$$\Psi_G(x) = \det(xI - L(G)) = \sum_{k=0}^{n} (-1)^{n-k} a_k x^k$$

and

$$Y_G(x) = \det(xI - S(G)) = \sum_{k=0}^{n} (-1)^{n-k} b_k x^k$$

as in Lemma 5.7.7, we can express IE(G) and LEL(G) as

$$\text{LEL}(G) = \frac{1}{2\pi} \int_0^\infty \frac{1}{\sqrt{u}} \ln \left(\sum_{k=0}^n a_k u^{k-n} \right) du$$

and

$$\text{IE}(G) = \frac{1}{2\pi} \int_0^\infty \frac{1}{\sqrt{u}} \ln \left(\sum_{k=0}^n b_k u^{k-n} \right) du,$$

see Theorem 5.7.9 and the discussion thereafter. Now the inequality follows from Lemma 5.7.7. For equality to hold, the characteristic polynomials must be identical. But then $L(G)$ and $S(G)$ have the same spectrum, which happens (by Corollary 5.2.3) if and only if G is bipartite. $\square$

We conclude this section with some extremal results and bounds. For trees, incidence energy and LEL coincide, so it suffices to consider one of the two. We can use the combinatorial interpretation of the coefficients of the characteristic polynomial of the Laplacian and the Coulson-type formula (5.17) to prove that the star and the path are extremal once again:

Theorem 5.7.13 *For every tree T with n vertices, we have*

$$n + \sqrt{n} - 2 = \text{LEL}(S_n) = \text{IE}(S_n) \le \text{LEL}(T) = \text{IE}(T)$$

and

$$\text{LEL}(T) = \text{IE}(T) \le \text{LEL}(P_n) = \text{IE}(P_n) = -1 + \cot \left(\frac{\pi}{4n} \right).$$

Equality holds in the first inequality if and only if T is the star, and equality holds in the second inequality if and only if T is the path.

Proof:

We already know that LEL(T) and IE(T) are always equal, since $L(T)$ and $S(T)$ have the same eigenvalues. The upper bound now follows easily by means of Theorem 5.7.8, which tells us that

$$\text{LEL}(T) = \text{IE}(T) = \frac{1}{2} \text{En}(\mathcal{S}(T)).$$

Since $\mathcal{S}(T)$ is again a tree, and only a path if T is, Theorem 5.5.1 gives us

$$\text{LEL}(T) = \text{IE}(T) = \frac{1}{2} \text{En}(\mathcal{S}(T)) \le \frac{1}{2} \text{En}(P_{2n-1})$$

$$= \frac{1}{2} \text{En}(\mathcal{S}(P_n)) = \text{LEL}(P_n) = \text{IE}(P_n),$$

with equality if and only if T is the path.

For the lower bound, we prove that S_n simultaneously minimizes all coefficients of the Laplacian characteristic polynomial (recall that these coefficients count rooted spanning forests), and that it is the only tree to do so. The statement then follows from the integral representation (5.17). So we need to show that

$$r_k(T) \geq r_k(S_n)$$

for all k, and that the only tree T for which equality holds for all k is the star S_n. We prove this by a direct counting argument. A rooted spanning forest of T with k components is obtained by removing $k - 1$ edges and picking a root for each of the resulting components. The number of choices for the edges to be removed is independent of the tree: it is always equal to $\binom{n-1}{k-1}$. For the star, removing $k - 1$ edges always yields $k - 1$ single leaves and a star with $n - k + 1$ vertices, so there are $n - k + 1$ choices of roots. Thus

$$r_k(S_n) = \binom{n-1}{k-1}(n - k + 1).$$

Now we consider an arbitrary tree T and remove $k - 1$ edges. If the resulting component sizes are $a_1, a_2, \ldots, a_k$, then the number of root choices is $a_1 a_2 \cdots a_k$. We also have $a_1 + a_2 + \cdots + a_k = n$. It is a standard exercise to show that the minimum of $a_1 a_2 \cdots a_k$ under this condition is $n - k + 1$, with equality if and only if all but one of the a_j are 1. Therefore, we have

$$r_k(T) \geq \binom{n-1}{k-1}(n - k + 1) = r_k(S_n)$$

for all k. It remains to show that equality can only hold for all k if T is the star. It suffices to prove this when $k = 2$: in this case, we know from the proof of Theorem 5.2.13 that $r_2(T)$ is equal to the Wiener index $W(T)$. Finally, we know from Proposition 2.3.2 that the star is the unique tree that minimizes the Wiener index. Hence

$$r_2(T) = W(T) > W(S_n) = r_2(S_n)$$

if T is not the star, completing the proof. $\square$

We remark that the lower bound remains valid for arbitrary connected graphs in view of the monotonicity properties (Lemmas 5.7.5 and 5.7.6): for every connected graph G, the inequalities $\mathrm{LEL}(S_n) \leq \mathrm{LEL}(G)$ and $\mathrm{IE}(S_n) \leq \mathrm{IE}(G)$ hold.

We can also provide bounds for incidence energy and Laplacian-energy-like invariant in terms of the number of vertices and/or edges and other parameters, in a similar way as for the graph energy. Let us just provide some basic examples.

Theorem 5.7.14 *For every graph G with n vertices and m edges, we have*

$$\mathrm{LEL}(G) \leq \sqrt{2m(n - 1)}.$$

Proof:

Let $\lambda_1, \lambda_2, \ldots, \lambda_n$ be the Laplacian eigenvalues, where $\lambda_n = 0$ without loss of generality. We apply the inequality between the arithmetic and the quadratic mean to obtain

$$\text{LEL}(G) = \sum_{i=1}^{n-1} \sqrt{\lambda_i} \leq \sqrt{(n-1) \sum_{i=1}^{n-1} \left(\sqrt{\lambda_i}\right)^2}.$$

Now observe that

$$\sum_{i=1}^{n-1} \left(\sqrt{\lambda_i}\right)^2 = \sum_{i=1}^{n-1} \lambda_i = \sum_{i=1}^{n} \lambda_i = \text{tr}(L(G)) = 2m$$

to complete the proof. □

Theorem 5.7.15 *For every graph G with n vertices and m edges, we have*

$$\text{LEL}(G) \geq \sqrt{2m}.$$

Proof:

Simply note that

$$\text{LEL}(G)^2 = \left(\sum_{i=1}^{n} \sqrt{\lambda_i}\right)^2 \geq \sum_{i=1}^{n} \left(\sqrt{\lambda_i}\right)^2 = \sum_{i=1}^{n} \lambda_i = \text{tr}(L(G)) = 2m.$$

□

We conclude this section with the remark that there are also further variants of the energy (such as skew energy, Randić energy, etc.) that are not covered here. We refer to [38] as a general reference.

5.8 Other invariants based on graph spectra

Among the many other important graph invariants that are based on graph spectra, we consider two more examples in this section: the spectral radius and the Estrada index.

5.8.1 Spectral radius of a graph

The spectral radius of any matrix is the greatest absolute value of an eigenvalue. The spectral radius of a graph G is simply the spectral radius of its adjacency matrix $A(G)$. We know that all eigenvalues of $A(G)$ are real, and the Perron-Frobenius Theorem on non-negative matrices (see Theorem 5.2.1) even guarantees that the spectral radius of $A(G)$ is an eigenvalue. The spectral radius, in the following denoted by $r(G)$, is certainly one of the most important quantities that have been studied in spectral graph theory. We refer to a recent book [100] devoted entirely to the topic for more information and results.

In Section 5.2, we related the spectrum of a graph to walks in the graph. The spectral radius occurs as a specific limit, which will be presented in the following proposition.

Proposition 5.8.1 *Let $W_r(G)$ denote the number of closed walks of length r in a graph G. If G is connected and not bipartite, then we have*

$$r(G) = \lim_{r \to \infty} \left(W_r(G)\right)^{1/r}.$$

If G is bipartite, then we have

$$r(G) = \lim_{r \to \infty} \left(W_{2r}(G)\right)^{1/(2r)}.$$

Proof:

Recall from Theorem 5.2.7 that

$$W_r(G) = \sum_{k=1}^{n} \alpha_k^r,$$

where $\alpha_1, \alpha_2, \ldots, \alpha_n$ are the eigenvalues of G. Without loss of generality, let $r(G) = \alpha_1$. If G is connected and not bipartite, then Theorem 5.2.3 guarantees that $-r(G)$ is not one of the eigenvalues. Thus the second factor in the expression

$$W_r(G) = \alpha_1^r \left(1 + \sum_{k=2}^{n} \left(\frac{\alpha_k}{\alpha_1}\right)^r\right)$$

goes to 1, since each of the fractions $\frac{\alpha_k}{\alpha_1}$ for $k > 1$ has absolute value less than 1. The statement of the proposition follows immediately. If G is bipartite, then $W_r(G) = 0$ for all odd r, so we restrict ourselves to even lengths and apply the same argument:

$$W_{2r}(G) = \alpha_1^{2r} \left(1 + \sum_{k=2}^{n} \left(\frac{\alpha_k}{\alpha_1}\right)^{2r}\right),$$

and the second factor clearly lies between 1 and n. Again, the statement of the proposition follows.   □

An important tool to analyze the spectral radius of graphs (and of symmetric matrices in general) is the Rayleigh quotient: for a symmetric matrix

M and a non-zero vector $\mathbf{x}$, set

$$R(M, \mathbf{x}) = \frac{\mathbf{x}^T M \mathbf{x}}{\mathbf{x}^T \mathbf{x}}.$$

If M is symmetric, then there exists an orthonormal basis of eigenvectors $\mathbf{v}_1, \mathbf{v}_2, \ldots, \mathbf{v}_n$ associated with the eigenvectors $\mu_1, \mu_2, \ldots, \mu_n$. We can write $\mathbf{x}$ as a linear combination of these eigenvectors:

$$\mathbf{x} = c_1 \mathbf{v}_1 + c_2 \mathbf{v}_2 + \cdots + c_n \mathbf{v}_n,$$

which gives us

$$\mathbf{x}^T M \mathbf{x} = \left(\sum_{i=1}^{n} c_i \mathbf{v}_i^T \right) \cdot M \cdot \left(\sum_{j=1}^{n} c_j \mathbf{v}_j^T \right)$$

$$= \sum_{i=1}^{n} \sum_{j=1}^{n} c_i c_j \mathbf{v}_i^T M \mathbf{v}_j$$

$$= \sum_{i=1}^{n} \sum_{j=1}^{n} c_i c_j \mu_j \mathbf{v}_i^T \mathbf{v}_j.$$

Since we are assuming that $\mathbf{v}_1, \mathbf{v}_2, \ldots, \mathbf{v}_n$ form an orthonormal basis, we have $\mathbf{v}_i^T \mathbf{v}_j = 0$ if $i \neq j$, and $\mathbf{v}_i^T \mathbf{v}_j = 1$ if $i = j$. Thus

$$\mathbf{x}^T M \mathbf{x} = \sum_{i=1}^{n} c_i^2 \mu_i.$$

By a similar calculation, we have

$$\mathbf{x}^T \mathbf{x} = \sum_{i=1}^{n} c_i^2.$$

Now it follows that

$$R(M, \mathbf{x}) = \frac{\sum_{i=1}^{n} c_i^2 \mu_i}{\sum_{i=1}^{n} c_i^2} \leq \max_i \mu_i,$$

with equality if and only if $c_i = 0$ for all indices i except for those for which μ_i is extremal (i.e., if $\mathbf{x}$ is an eigenvector for the largest eigenvalue). This gives us the following important characterization of the largest eigenvalue:

$$\max_i \mu_i = \max_{\mathbf{x} \neq \mathbf{0}} R(M, \mathbf{x}).$$

In particular, every Rayleigh quotient $R(M, \mathbf{x})$ provides a lower bound on the maximum eigenvalue and thus the spectral radius. This is very useful to derive bounds on the spectral radius. As a first application, we show that it is monotone with respect to adding/removing vertices and edges:

Proposition 5.8.2 *If H is obtained from G by removing a vertex or an edge, then we have*

$$r(H) \leq r(G).$$

If G is connected, then the inequality is strict.

Proof:

Consider first the case that H is obtained by removing a vertex. Let $\mathbf{x}$ be an eigenvector corresponding to $r(H)$ with non-negative entries, as guaranteed by the Perron-Frobenius Theorem (Theorem 5.2.1). We define a new vector $\mathbf{x}'$ whose entries are the same as those of $\mathbf{x}$, with an additional zero entry corresponding to the vertex that is part of G, but not H. It is easy to see that

$$R(A(G), \mathbf{x}') = R(A(H), \mathbf{x}),$$

hence

$$r(G) \geq R(A(G), \mathbf{x}') = R(A(H), \mathbf{x}) = r(H).$$

Now consider the case that H is obtained by removing an edge. Take $\mathbf{x}$ as before, and observe that

$$r(G) \geq R(A(G), \mathbf{x}) \geq R(A(H), \mathbf{x}) = r(H),$$

since the entries of $A(G)$ are greater than or equal to those of $A(H)$. This completes the proof of the inequality. In the first case (vertex removal), equality can only hold if $\mathbf{x}'$ and $\mathbf{x}$ are both eigenvectors for G and H, respectively. If G is connected, this is impossible since the entries of $\mathbf{x}'$ would all have to be positive by the Perron-Frobenius Theorem. In the second case (edge removal), $\mathbf{x}$ has to be an eigenvector for both G and H (with respect to the eigenvalue $r(G) = r(H)$) for equality to hold. If G is connected, $\mathbf{x}$ has to have positive entries, so $A(G)\mathbf{x} \neq A(H)\mathbf{x}$. This gives another contradiction, so equality cannot hold in this case either.   □

We remark that the statement of Proposition 5.8.2 regarding the removal of vertices also follows from Corollary 5.2.8 (interlacing property). Next, we prove some elementary bounds on the spectral radius in terms of degrees.

Theorem 5.8.1 *Let G be a graph with n vertices and m edges. Let $\bar{d} = \frac{2m}{n}$ denote its average degree, and Δ its maximum degree. The spectral radius $r(G)$ satisfies*

$$\bar{d} \leq r(G) \leq \Delta.$$

The lower bound holds with equality if and only if G is regular, the upper bound holds with equality if and only if at least one connected component of G is Δ-regular.

Proof:

For the lower bound, we consider the Rayleigh quotient of the vector $\mathbf{1}$ whose entries are all 1s. It follows from the definition of the adjacency matrix that the entries of $A(G) \cdot \mathbf{1}$ are precisely the degrees of the vertices. Thus

$$\mathbf{1}^T A(G)\mathbf{1} = \sum_{v \in V(G)} \deg(v) = 2m,$$

and since $\mathbf{1}^T\mathbf{1} = n$, it follows that

$$r(G) \geq R(A(G), \mathbf{1}) = \frac{2m}{n} = \bar{d}.$$

For equality to hold, $\mathbf{1}$ has to be an eigenvector corresponding to $r(G)$. This happens if and only if all entries of $A(G) \cdot \mathbf{1}$ are equal, i.e., all vertex degrees are the same (if $\mathbf{1}$ is an eigenvector, it has to be the eigenvector associated with the spectral radius by the Perron-Frobenius Theorem).

For the upper bound, let $\mathbf{x}$ be an eigenvector with positive entries corresponding to the spectral radius $r(G)$. Let $x_1, x_2, \ldots, x_n$ be the entries of this eigenvector, corresponding to vertices $v_1, v_2, \ldots, v_n$, and choose i in such a way that x_i is maximal. The eigenvalue equation $r(G)\mathbf{x} = A(G)\mathbf{x}$ gives us

$$r(G)x_i = \sum_{j:v_j \in N(v_i)} x_j \leq \deg(v_i)x_i \leq \Delta x_i, \tag{5.21}$$

thus $r(G) \leq \Delta$. Let us now analyze the cases of equality: the only way that equality can hold in (5.21) is that $\deg(v_i) = \Delta$ and $x_j = x_i$ for all vertices v_j that are adjacent to v_i. Iterating this argument, we find that the degrees of all vertices in the same connected component have to be equal to Δ. Conversely, if a connected component is Δ, then the all-ones vector is an eigenvector for that component, corresponding to the eigenvalue Δ. This completes the proof of the theorem. $\qquad\square$

In view of Proposition 5.8.2, we expect that the edgeless graph and the complete graph are extremal, which is the statement of the following corollary:

Corollary 5.8.1 *For every graph G with n vertices, we have*

$$0 \leq r(G) \leq n - 1.$$

Equality holds for the lower bound if and only if G is edgeless, and equality holds for the upper bound if and only if G is the complete graph.

Proof:

This follows easily from the previous theorem, noting that

$$0 \leq \bar{d} \leq \Delta \leq n - 1.$$

Moreover, the only 0-regular graph with n vertices is the edgeless graph, and the only graph with n vertices and an $(n-1)$-regular component is the complete graph. $\qquad\square$

It is natural to expect that the star and the path are extremal among trees, and this is indeed the case. We first prove this for the star.

Theorem 5.8.2 *For every tree T with n vertices, we have*

$$r(T) \le r(S_n) = \sqrt{n-1},$$

with equality if and only if T is a star.

Proof:

Let the eigenvalues of T be $\alpha_1, \alpha_2, \ldots, \alpha_n$. Since trees are bipartite, we know that $r(T)$ and $-r(T)$ are both eigenvalues by Theorem 5.2.2, say $\alpha_1 = r(T)$ and $\alpha_2 = -r(T)$. Moreover, Corollary 5.2.4 yields

$$\sum_{i=1}^{n} \alpha_i^2 = 2(n-1),$$

so

$$2r(T)^2 = \alpha_1^2 + \alpha_2^2 \le \sum_{i=1}^{n} \alpha_i^2 = 2(n-1),$$

which implies that $r(T) \le \sqrt{n-1}$. Equality holds if and only if the spectrum consists of $\pm\sqrt{n-1}$ and 0 (with multiplicity $n-2$). Equivalently, the characteristic polynomial has to be $x^n - (n-1)x^{n-2}$. This is the case for the star, and it is the only such tree: in view of Corollary 5.2.5, $m(T,2)$ needs to be 0 to yield this characteristic polynomial, and the star is clearly the only tree without matchings of cardinality 2. $\qquad\square$

The analogous statement for the minimum is somewhat more complicated. We use an approach due to Lovász and Pelikán [74] that is based on the following lemmas:

Lemma 5.8.1 *Let F be a forest, and let G be obtained from F by removing one or more edges. For every $x \ge r(F)$, the following inequality for the characteristic polynomials holds:*

$$\Phi_F(x) \le \Phi_G(x).$$

Strict inequality always holds for $x > r(F)$.

Proof:

It suffices to prove the statement for the case where G is obtained by removing a single edge e; the general statement follows by induction. Recall that the characteristic polynomials of F and G coincide with their matching polynomials by Corollary 5.2.5. Hence, we can use the recursions of Lemma 4.7.2. Specifically, if v and w are the ends of edge e, we have

$$\Phi_F(x) = M(F; x) = M(F - e; x) - M(F - \{v, w\}; x) = \Phi_G(x) - \Phi_{F-\{v,w\}}(x).$$

By Proposition 5.8.2, we know that the spectral radius of $F - \{v, w\}$ is less than or equal to the spectral radius of F, so $x \geq r(F) \geq r(F - \{v, w\})$. But this means that $\Phi_{F-\{v,w\}}(x) \geq 0$, since x is greater than or equal to all the zeros of the polynomial $\Phi_{F-\{v,w\}}(x)$, and the leading coefficient of this polynomial is positive. For $x > r(F)$, we even have $\Phi_{F-\{v,w\}}(x) > 0$ by the same argument. □

The following lemma is somewhat similar to (a special case of) Lemma 4.4.3.

Lemma 5.8.2 *Let v be a vertex of a tree T with more than one vertex. Let T_1 be obtained from T by appending paths of length k and ℓ ($k, \ell \geq 1$) to v, and let T_2 be obtained from T by appending a single path of length $k + \ell$ to v. Then we have*

$$r(T_1) > r(T_2).$$

Proof:

We will prove the stronger statement that

$$\Phi_{T_2}(x) > \Phi_{T_1}(x) \geq 0$$

for $x \geq r(T_1)$. This implies that there is no zero of Φ_{T_2} greater than or equal to $r(T_1)$, which proves the lemma. The inequality $\Phi_{T_1}(x) \geq 0$ is clear, since $r(T_1)$ is the largest zero and the leading coefficient of Φ_{T_1} is positive. For the inequality $\Phi_{T_2}(x) > \Phi_{T_1}(x)$, we make use of the fact that Φ_{T_1} and Φ_{T_2} are equal to the matching polynomials of T_1 and T_2, respectively, as in the proof of the previous lemma. Let w be a neighbor of v on the path of length k in T, and let e be the edge between v and w. In T_2, we let e' be an edge on the path of length $k + \ell$ whose removal divides the path into a part of length ℓ (with v at one of its ends) and a path of length $k - 1$. Let v' and w' be the ends of e' (v' being the vertex that is closer to v). Note that $T_1 - e$ and $T_2 - e'$ are isomorphic: both consist of T with a path of length ℓ attached to v, and a path of length $k - 1$. The recursions of Lemma 4.7.2 yield

$$\Phi_{T_1}(x) = M(T_1; x) = M(T_1 - e; x) - M(T_1 - \{v, w\}; x)$$

and

$$\Phi_{T_2}(x) = M(T_2; x) = M(T_2 - e'; x) - M(T_2 - \{v', w'\}; x),$$

so since $M(T_1 - e; x) = M(T_2 - e'; x)$, we have to prove that

$$M(T_1 - \{v, w\}; x) = \Phi_{T_1 - \{v,w\}}(x)$$
$$> \Phi_{T_2 - \{v',w'\}}(x) = M(T_2 - \{v', w'\}; x) \quad (5.22)$$

for all $x \geq r(T_1)$. Now note that a graph isomorphic to $T_1 - \{v, w\}$ (which consists of paths of length $k - 2$ and $\ell - 1$ and $T - v$) can be obtained by removing edges from $T_2 - \{v', w'\}$ (which consists of a path of length $\ell - 1$ and T with a path of length $k - 2$ attached to v). Moreover, $x \geq r(T_1) > r(T_2 - \{v', w'\})$ by Proposition 5.8.2 (since $T_2 - \{v', w'\}$ is isomorphic to a subgraph of T_1), so (5.22) holds by Lemma 5.8.1. This completes the proof. $\square$

Theorem 5.8.3 *For every tree T with n vertices, we have*

$$r(T) \geq r(P_n) = 2 \cos \frac{\pi}{n + 1},$$

with equality if and only if T is a path.

Proof:

The formula for the spectral radius of the path follows from Proposition 5.2.4. Now suppose that T is a tree with n vertices for which the spectral radius attains its minimum. If T is not a path, then it contains vertices of degree 3 or higher. Consider the smallest subtree S of T that contains all these vertices (possibly S is just a single vertex), and let v be a leaf of S. This vertex has to have degree 3 or more in T, since it would otherwise not be contained in S. Moreover, since v is a leaf of S, at most one of the branches at v contains other vertices of degree 3 or more. Thus there are at least two branches that are paths (with one end at v). Pick two such paths, and suppose that their lengths are k and ℓ, respectively. If we replace them by a single path of length $k + \ell$, then the spectral radius decreases by Lemma 5.8.2. Since this contradicts our choice of T, the path must be the only tree for which the spectral radius attains its minimum. $\square$

It is clear that $r(G) \leq \text{En}(G)$ for all graphs G, since the energy is the sum of the absolute values of all eigenvalues, which includes $r(G)$. This immediately gives us bounds like the following:

Corollary 5.8.2 *For every graph G with n vertices and m edges, we have*

$$r(G) \leq \sqrt{2mn}.$$

Proof:

This follows from the aforementioned inequality $r(G) \leq \text{En}(G)$, combined with Theorem 5.4.1. $\square$

Since the inequality $r(G) \leq \text{En}(G)$ is clearly not optimal, the bound in terms of the number of vertices and edges can be improved considerably. This is done in the following theorem:

Theorem 5.8.4 *For every graph G with n vertices and m edges, we have*

$$r(G) \leq \sqrt{\frac{2m(n-1)}{n}}.$$

Proof:

Let $\alpha_1 = r(G), \alpha_2, \ldots, \alpha_n$ be the eigenvalues. The proof is based on the two identities of Corollary 5.2.4:

$$\sum_{i=1}^{n} \alpha_i = 0$$

and

$$\sum_{i=1}^{n} \alpha_i^2 = 2m.$$

We clearly have

$$\alpha_1 = -\sum_{i=2}^{n} \alpha_i \leq \sum_{i=2}^{n} |\alpha_i|.$$

The inequality between the arithmetic and the quadratic mean gives us

$$\frac{1}{n-1} \sum_{i=2}^{n} |\alpha_i| \leq \sqrt{\frac{1}{n-1} \sum_{i=2}^{n} \alpha_i^2} = \sqrt{\frac{2m - \alpha_1^2}{n-1}},$$

so

$$\frac{\alpha_1}{n-1} \leq \sqrt{\frac{2m - \alpha_1^2}{n-1}}.$$

Elementary manipulations now yield

$$r(G) = \alpha_1 \leq \sqrt{\frac{2m(n-1)}{n}}.$$

$\square$

We conclude this section with a surprising relation between the chromatic number of a graph (the smallest number of colors needed to color each vertex in such a way that adjacent vertices are not assigned the same color).

Theorem 5.8.5 *Let $\chi(G)$ be the chromatic number of a graph G. We have*

$$\chi(G) \leq r(G) + 1.$$

Proof:

Write k for the chromatic number of G, and let H be an induced subgraph of G with the smallest number of vertices such that $\chi(H) = \chi(G) = k$. Consider an arbitrary vertex v of H. Since $H - v$ has smaller chromatic number than H by our choice of v, it can be colored with $k - 1$ colors in such a way that no two adjacent vertices are assigned the same color. If the degree of v in H is less than $k - 1$, then there are at most $k - 2$ distinct colors among the neighbors of v in H, so v can be colored with one of the $k - 1$ colors used to color $H - v$ to obtain a feasible coloring of H with $k - 1$ colors, contradicting the assumption that $\chi(H) = k$. Thus the degree of v is at least $k - 1$. Since this holds for every vertex of H, the average degree of H is also at least $k - 1$. Now Theorem 5.8.1 implies that

$$\chi(G) - 1 = k - 1 \leq r(H),$$

and since $r(H) \leq r(G)$ in view of Proposition 5.8.2, the statement of the theorem follows. □

5.8.2   Estrada index

The Estrada index was put forward by Estrada [27]. Similar to the graph energy, it is defined as a sum over all eigenvalues. Specifically, it is given by

$$EE(G) = \sum_{i=1}^{n} e^{\alpha_i},$$

where $\alpha_1, \alpha_2, \ldots, \alpha_n$ are the eigenvalues (of the adjacency matrix) of G. An alternative expression makes use of the connection between walks and the spectrum of the adjacency matrix:

Proposition 5.8.3 *Let $W_r(G)$ denote the number of closed walks of length r in G. The Estrada index of G can be expressed as*

$$EE(G) = \sum_{r=0}^{\infty} \frac{W_r(G)}{r!}.$$

Proof:

By Theorem 5.2.7, the total number $W_r(G)$ of closed walks of length r in G is equal to the trace of $A(G)^r$:

$$W_r(G) = \mathrm{tr}(A(G)^r) = \sum_{i=1}^{n} \alpha_i^r.$$

Combining this with the series expansion of the exponential function, we obtain

$$EE(G) = \sum_{i=1}^{n} e^{\alpha_i} = \sum_{i=1}^{n} \sum_{r=0}^{\infty} \frac{\alpha_i^r}{r!} = \sum_{r=0}^{\infty} \frac{W_r(G)}{r!},$$

which is exactly the identity we wanted to prove. □

The representation in terms of the number of closed walks immediately shows that the Estrada index is monotone with respect to vertex and edge addition/removal:

Proposition 5.8.4 *If H is obtained from G by removing a vertex or an edge, then we have*
$$\mathrm{EE}(H) < \mathrm{EE}(G).$$

Proof:

Clearly, every closed walk of H is also a closed walk of G, and there are some closed walks in G that are not in H (using the additional vertex or edge). Hence we have $W_r(H) \leq W_r(G)$ for all r, and the inequality is strict for some r. The desired inequality now follows from Proposition 5.8.3. □

The following corollary is immediate:

Corollary 5.8.3 *For every graph G with n vertices, we have*
$$n = \mathrm{EE}(E_n) \leq \mathrm{EE}(G),$$
with equality if and only if G is the edgeless graph, and
$$e^{n-1} + (n-1)e^{-1} = \mathrm{EE}(K_n) \geq \mathrm{EE}(G),$$
with equality if and only if G is the complete graph.

Once again, the star and the path are extremal among trees (and in view of Proposition 5.8.4, the path even attains the minimum Estrada index among all connected graphs). This will be shown in the following, based on the work of Deng [22]. Graph transformations similar to those applied, e.g., in the analysis of Merrifield-Simmons index, Hosoya index and spectral radius will be important tools, and the expression for the Estrada index in terms of closed walks will play a major role again.

Lemma 5.8.3 *Let v be the center of a star with n vertices, and u a leaf. The number of closed walks of length r starting and ending at v is $(n-1)^{r/2}$ for even r and 0 otherwise. The number of closed walks of length r starting and ending at u is 1 for $r = 0$, $(n-1)^{r/2-1}$ for even $r > 0$, and 0 otherwise.*

Proof:

A closed walk starting at v begins with a move to one of the $n-1$ leaves, and a return to v. This is repeated $r/2$ times for even r, so there are $(n-1)^{r/2}$ possibilities. Since all walks alternate between the center and the leaves, there are no closed walks of odd length. The proof for walks starting and ending at u is analogous, except for the fact that there is no choice for the first and last step. □

Lemma 5.8.4 *Let w be an arbitrary vertex of a tree T. Let T_1 be obtained from T by merging w with the center of an n-vertex star $(n \geq 3)$, and let T_2 be obtained by merging w with a leaf of an n-vertex star. For every $r \geq 0$, we have*

$$W_r(T_1) \geq W_r(T_2),$$

and the inequality is strict for all even $r \geq 4$.

Proof:

Let v and u be center and leaf of the star that are merged with w, as in the previous lemma. Closed walks of T_1 and T_2 can be divided into three categories: those that stay within T, those that stay within the star, and those that use edges of both. Clearly, there are equally many walks of the first two types in T_1 and T_2, so we focus on the third type. We define an injection Ξ from walks of the third type in T_2 to walks of the third type in T_1.

By the previous lemma, there is an injection ξ from the set of closed paths of a given length starting at v to the set of closed paths of the same length starting at u, and this injection is not surjective for closed walks of even length greater than or equal to 2. Every closed walk in T_2 that uses edges of both T and the star can be decomposed by splitting it at all those times when w is reached. The parts of this decomposition are the maximal subwalks that either only use edges of T or only use edges of the star, and they are all closed walks starting and ending at w, except possibly for the first and last part (if the walk does not start at w), which however also form a walk starting and ending at w when they are joined.

We define the injection Ξ in the following simple way: all parts that use only edges of T are kept as they are, while parts that use edges of the star are replaced by their images under the injection ξ. Clearly, this is again an injection, and it is not surjective for walks of even length greater than or equal to 4: to see why this is the case, consider any of the closed walks of length 2 in the star starting and ending at the center v that does not have a preimage under ξ. Append this walk to an arbitrary closed walk of length $r - 2$ in T that starts and ends at w to obtain a closed walk in T_1 that does not have a preimage under Ξ. This proves the inequality

$$W_r(T_1) \geq W_r(T_2),$$

with strict inequality if r is even and $r \geq 4$. $\square$

Theorem 5.8.6 *For every tree T with n vertices, we have*

$$\mathrm{EE}(T) \leq \mathrm{EE}(S_n) = e^{\sqrt{n-1}} + e^{-\sqrt{n-1}} + n - 2,$$

and equality holds if and only if T is a star.

Proof:

Let T be a tree with n vertices that attains the maximum Estrada index. If T is not a star, consider any diametrical path (path of greatest length) $v_0, v_1, \ldots, v_k$. Then all but one of the neighbors of v_{k-1} (namely v_{k-2}) are leaves, since otherwise there would be a way to extend the path for at least two edges beyond v_{k-1}, giving a longer path. Hence the tree induced by v_{k-1} and its neighbors is a star, and v_{k-2} is one of its leaves. Thus we are in the situation of Lemma 5.8.4 (v_{k-2} taking the role of w), and we are able to find another tree that has at least as many closed walks of every even length as T, and strictly more for all even lengths greater than or equal to 4. In view of Proposition 5.8.3, this tree would have greater Estrada index than T, which is a contradiction. Hence T has to be a star. The formula for the Estrada index of S_n is immediate from Corollary 5.2.1. $\qquad \square$

Lemma 5.8.5 *Let u be a leaf of a path with n vertices, and let v be a non-leaf of the same path. There is an injective function from the set of closed walks of length r starting and ending at u to the set of closed walks of length r starting and ending at v. If r is even and greater than or equal to 2, then this function is not surjective.*

Proof:

Let the vertices of the path be $w_1, w_2, \ldots, w_n$ in this order, with $u = w_1$ and $v = w_k$ for some $k \subset \{2, 3, \ldots, n-1\}$. Let ψ be the "mirror" map on the set $\{w_1, w_2, \ldots, w_k\}$, defined by $\psi(w_r) = w_{k+1-r}$. This map also induces a bijection between the set of closed walks in the path induced by $\{w_1, w_2, \ldots, w_k\}$ that start and end at $u = w_1$ and those that start and end at $v = w_k$. Now we define an injection Ψ that maps closed walks starting and ending at u to those that start and end at v. There are two possible cases for a closed path W that starts and ends at u:

- If the walk does not reach v, then it lies entirely inside of the subpath induced by $\{w_1, w_2, \ldots, w_k\}$. In this case, we simply apply the mirror map ψ to obtain the image of W under Ψ, which is a closed walk starting and ending at v.

- Otherwise, we decompose the walk into two pieces: the piece W_1 up to the first time v is reached, and the rest, which we denote by W_2. The image $\Psi(W)$ is obtained as follows: start with W_2, and append the image of W_1 under ψ, but with directions reversed (so that first and last vertex are interchanged). Given $\Psi(W)$, the decomposition is unique: W_2 corresponds to the part up to the last time that the walk reaches u. Note here that since W_1 only contains v as its final vertex, its image under ψ only contains u as its final vertex, so the reversed image only contains it as its initial vertex.

In both cases, the preimage of Ψ is unique, and since the images in the first

244 *Introduction to Chemical Graph Theory*

case never reach the vertex u while they do in the second case, we cannot get identical images under Ψ from the two cases either. Thus Ψ is an injection. To see why it is not a surjection except for the trivial cases where the length is 0 or odd, consider any closed walk starting and ending at v that does not reach u, and does not lie entirely inside of the subpath induced by $\{w_1, w_2, \ldots, w_k\}$. Such a path does not have a preimage under Ψ. □

Lemma 5.8.6 *Let w be an arbitrary vertex of a tree T. Let T_1 be obtained from T by appending paths of length k and ℓ ($k, \ell \geq 1$) to w, and let T_2 be obtained from T by appending a single path of length $k + \ell$ to w. For every $r \geq 0$, we have*

$$W_r(T_1) \geq W_r(T_2),$$

and the inequality is strict for all even $r \geq 4$.

Proof:

Analogous to the proof of Lemma 5.8.4, using Lemma 5.8.5 instead of Lemma 5.8.3. □

Theorem 5.8.7 *For every tree T with n vertices, we have*

$$\mathrm{EE}(T) \geq \mathrm{EE}(P_n),$$

and equality holds if and only if T is a path.

Proof:

This theorem follows in the same way as Theorem 5.8.3, using Lemma 5.8.6 instead of Lemma 5.8.2. □

Exercises

1. Let M be a symmetric $n \times n$ matrix. Prove that the k-th largest eigenvalue of M is given by

$$\max_{\dim(V)=k} \min_{\mathbf{x}\in V\setminus\{\mathbf{0}\}} R(M, \mathbf{x}),$$

where the first maximum is over all k-dimensional subspaces of $\mathbb{R}^n$. Likewise, prove that the k-th smallest eigenvalue of M is given by

$$\min_{\dim(V)=k} \max_{\mathbf{x}\in V\setminus\{\mathbf{0}\}} R(M, \mathbf{x}).$$

Deduce the Cauchy interlacing theorem (Theorem 5.2.10) from these formulas.

2. Prove that a tree T has a perfect matching (a matching that covers all vertices) if and only if 0 is not an eigenvalue of T.

3. Let $\alpha_1, \alpha_2, \ldots, \alpha_n$ be the eigenvalues of a tree T. Prove the following formula for the Hosoya index of T:

$$Z(T) = \prod_{i=1}^{n} \sqrt{1 + \alpha_i^2}.$$

4. Applying Theorem 5.2.14 to a tree T with n vertices, we see that the number of rooted spanning forests with k components in T is equal to the number of matchings of the subdivision graph $\mathcal{S}(T)$ with $n - k$ edges. Find a bijective proof for this fact.

5. Prove the asymptotic formulas in (5.5).

6. Prove Theorem 5.7.3 and Theorem 5.7.4.

7. Let G be a graph with n vertices, m edges and p connected components. Prove that

$$\text{LEL}(G) \leq \sqrt{2m(n-p)}.$$

8. Prove the following statement from the proof of Theorem 5.7.13: the minimum of the product $a_1 a_2 \cdots a_k$ for positive integers $a_1, a_2, \ldots, a_k$ satisfying $a_1 + a_2 + \cdots + a_k = n$ is $n - k + 1$.

9. Prove the following inequalities for the incidence energy:

$$\sqrt{2m} \leq \text{IE}(G) \leq \sqrt{2mn}$$

for all graphs G with n vertices and m edges.

10. Let v_i and v_j be vertices of a non-bipartite connected graph G, and let $w_{ij}^{(r)}$ denote the number of walks of length r from v_i to v_j. Prove that

$$\lim_{r \to \infty} \left(w_{ij}^{(r)} \right)^{1/r} = r(G).$$

11. Prove the following inequality for the spectral radius: if Δ is the maximum degree of a graph G, then

$$r(G) \geq \sqrt{\Delta}.$$

12. Deduce the inequalities

$$r(P_n) \leq r(T) \leq r(S_n)$$

for all trees T with n vertices from Lemma 5.8.4 and Lemma 5.8.6.

13. Prove that

$$\lim_{n \to \infty} \frac{1}{n} \text{EE}(P_n) = \frac{1}{\pi} \int_0^\pi e^{2 \cos t} \, dt.$$

Bibliography

[1] J. Aihara. A new definition of Dewar–type resonance energies. *J. Am. Chem. Soc.*, 98:2750–2758, 1976.

[2] E. O. D. Andriantiana. Unicyclic bipartite graphs with maximum energy. *MATCH Commun. Math. Comput. Chem.*, 66(3):913–926, 2011.

[3] E. O. D. Andriantiana. Energy, Hosoya index and Merrifield-Simmons index of trees with prescribed degree sequence. *Discrete Appl. Math.*, 161(6):724–741, 2013.

[4] E. O. D. Andriantiana and S. Wagner. Unicyclic graphs with large energy. *Linear Algebra Appl.*, 435(6):1399–1414, 2011.

[5] E. O. D. Andriantiana, S. Wagner, and H. Wang. Maximum Wiener index of trees with given segment sequence. *MATCH Commun. Math. Comput. Chem.*, 75(1):91–104, 2016.

[6] M. Bartlett, E. Krop, C. Magnant, F. Mutiso, and H. Wang. Variations of distance-based invariants of trees. *J. Combin. Math. Combin. Comput.*, 91:19–29, 2014.

[7] N. L. Biggs, E. K. Lloyd, and R. J. Wilson. *Graph theory: 1736–1936*. Clarendon Press, Oxford, 1976.

[8] B. Bollobás and P. Erdős. Graphs of extremal weights. *Ars Combin.*, 50:225–233, 1998.

[9] B. Bollobás, P. Erdős, and A. Sarkar. Extremal graphs for weights. *Discrete Math.*, 200(1-3):5–19, 1999. Paul Erdős memorial collection.

[10] A. E. Brouwer and W. H. Haemers. *Spectra of graphs*. Universitext. Springer, New York, 2012.

[11] G. Caporossi, D. Cvetković, I. Gutman, and P. Hansen. Variable Neighborhood Search for Extremal Graphs. 2. Finding Graphs with Extremal Energy. *J. Chem. Inf. Comput. Sci.*, 39(3):984–996, 1999.

[12] G. Caporossi, I. Gutman, P. Hansen, and L. Pavlović. Graphs with maximum connectivity index. *Comput. Biol. Chem.*, 27:85–90, 2000.

[13] A. Cayley. On the mathematical theory of isomers. *Philos. Mag.*, 47:444–446, 1874.

[14] E. Çela, N. S. Schmuck, S. Wimer, and G. J. Woeginger. The Wiener maximum quadratic assignment problem. *Discrete Optim.*, 8(3):411–416, 2011.

[15] Y.-H. Chen, H. Wang, and X.-D. Zhang. Note on extremal graphs with given matching number. *Appl. Math. Comput.*, 308:149–156, 2017.

[16] M. Chudnovsky and P. Seymour. The roots of the independence polynomial of a clawfree graph. *J. Combin. Theory Ser. B*, 97(3):350–357, 2007.

[17] A. Collins, F. Mutiso, and H. Wang. Optimal trees for functions of internal distance. *Involve*, 5(3):371–378, 2012.

[18] C. A. Coulson. On the calculation of the energy in unsaturated hydrocarbon molecules. *Proc. Cambridge Phil. Soc.*, 36:201–203, 1986.

[19] D. M. Cvetković, M. Doob, and H. Sachs. *Spectra of graphs – Theory and application*. Johann Ambrosius Barth, Heidelberg, third edition, 1995.

[20] P. Dankelmann, W. Goddard, and C. S. Swart. The average eccentricity of a graph and its subgraphs. *Util. Math.*, 65:41–51, 2004.

[21] K. Das, I. Gutman, and B. Furtula. On atom-bond connectivity index. *Chem. Phys. Lett.*, 511:452–454, 2011.

[22] H. Deng. A proof of a conjecture on the Estrada index. *MATCH Commun. Math. Comput. Chem.*, 62(3):599–606, 2009.

[23] H. Deng and S. Chen. The extremal unicyclic graphs with respect to Hosoya index and Merrifield-Simmons index. *MATCH Commun. Math. Comput. Chem.*, 59(1):171–190, 2008.

[24] A. A. Dobrynin, R. Entringer, and I. Gutman. Wiener index of trees: theory and applications. *Acta Appl. Math.*, 66(3):211–249, 2001.

[25] R. C. Entringer, D. E. Jackson, and D. A. Snyder. Distance in graphs. *Czechoslovak Math. J.*, 26(101)(2):283–296, 1976.

[26] P. Erdős and T. Gallai. Gráfok előírt fokszámú pontokkal. *Matematikai Lapok*, 11:264–274, 1960.

[27] E. Estrada. Characterization of 3d molecular structure. *Chem. Phys. Lett.*, 319:713–718, 2000.

[28] E. Estrada, L. Torres, L. Rodríguez, and I. Gutman. An atom-bond connectivity index: modelling the enthalpy of formation of alkanes. *Indian J. Chem.*, 37(10):849–855, 1998.

[29] K. Fan. Maximum properties and inequalities for the eigenvalues of completely continuous operators. *Proc. Nat. Acad. Sci., U. S. A.*, 37:760–766, 1951.

[30] M. Fischermann, A. Hoffmann, D. Rautenbach, L. Székely, and L. Volkmann. Wiener index versus maximum degree in trees. *Discrete Appl. Math.*, 122(1-3):127–137, 2002.

[31] E. Fritscher, C. Hoppen, I. Rocha, and V. Trevisan. On the sum of the Laplacian eigenvalues of a tree. *Linear Algebra Appl.*, 435(2):371–399, 2011.

[32] I. Gutman. Acyclic systems with extremal Hückel pi-electron energy. *Theor. Chim. Acta*, 45:79–87, 1977.

[33] I. Gutman. The energy of a graph. *Ber. Math.-Statist. Sekt. Forsch. Graz*, (100-105):Ber. No. 103, 22, 1978. 10. Steiermärkisches Mathematisches Symposium (Stift Rein, Graz, 1978).

[34] I. Gutman. Graphs with greatest number of matchings. *Publ. Inst. Math. (Beograd) (N.S.)*, 27(41):67–76, 1980.

[35] I. Gutman. Graphs with maximum and minimum independence numbers. *Publ. Inst. Math. (Beograd) (N.S.)*, 34(48):73–79, 1983.

[36] I. Gutman, B. Furtula, and M. Petrović. Terminal Wiener index. *J. Math. Chem.*, 46(2):522–531, 2009.

[37] I. Gutman, D. Kiani, M. Mirzakhah, and B. Zhou. On incidence energy of a graph. *Linear Algebra Appl.*, 431(8):1223–1233, 2009.

[38] I. Gutman and X. Li, editors. *Energies of Graphs – Theory and Applications*, volume 17 of *Mathematical Chemistry Monographs*. 2016.

[39] I. Gutman, S. Marković, and Ž. Bančević. Correlation between Wiener and quasi-Wiener indices in benzenoid hydrocarbons. *J. Serb. Chem. Soc.*, 60:633–636, 1995.

[40] I. Gutman, O. Miljković, G. Caporossi, and P. Hansen. Alkanes with small and large Randić connectivity indices. *Chem. Phys. Letters*, 306(5):366–372, 1999.

[41] I. Gutman, M. Milun, and N. Trinajstić. Topological definition of delocalisation energy. *MATCH Commun. Math. Comput. Chem.*, 1:171–175, 1975.

[42] I. Gutman, M. Milun, and N. Trinajstić. Graph theory and molecular orbitals. 19. Nonparametric resonance energies of arbitrary conjugated systems. *J. Am. Chem. Soc.*, 99:1692–1704, 1977.

[43] I. Gutman and B. Mohar. The quasi-Wiener and the Kirchhoff indices coincide. *J. Chem. Inf. Comput. Sci.*, 36:982–985, 1996.

[44] I. Gutman and N. Trinajstić. Graph theory and molecular orbitals. Total π-electron energy of alternant hydrocarbons. *Chem. Phys. Lett.*, 17:535–538, 1972.

[45] I. Gutman and S. Wagner. The matching energy of a graph. *Discrete Appl. Math.*, 160(15):2177–2187, 2012.

[46] I. Gutman and Y.-N. Yeh. The sum of all distances in bipartite graphs. *Math. Slovaca*, 45(4):327–334, 1995.

[47] I. Gutman and B. Zhou. Laplacian energy of a graph. *Linear Algebra Appl.*, 414(1):29–37, 2006.

[48] O. J. Heilmann and E. H. Lieb. Theory of monomer-dimer systems. *Comm. Math. Phys.*, 25:190–232, 1972.

[49] H. Hosoya. Topological index. A newly proposed quantity characterizing the topological nature of structural isomers of saturated hydrocarbons. *Bull. Chem. Soc. Japan*, 44(9):2332–2339, 1971.

[50] H. Hosoya. Topological index as a common tool for quantum chemistry, statistical mechanics, and graph theory. In *Mathematical and computational concepts in chemistry (Dubrovnik, 1985)*, Ellis Horwood Ser. Math. Appl., pages 110–123. Horwood, Chichester, 1986.

[51] H. Hosoya. The Topological Index Z Before and After 1971. *Internet Electron. J. Mol. Des.*, 1:428–442, 2002.

[52] Y. Hou. Unicyclic graphs with minimal energy. *J. Math. Chem.*, 29(3):163–168, 2001.

[53] Y. Hou, I. Gutman, and C.-W. Woo. Unicyclic graphs with maximal energy. *Linear Algebra Appl.*, 356:27–36, 2002. Special issue on algebraic graph theory (Edinburgh, 2001).

[54] Y. Hu, X. Li, and Y. Yuan. Trees with minimum general Randić index. *MATCH Commun. Math. Comput. Chem.*, (52):119–128, 2004.

[55] H. Hua. Bipartite unicyclic graphs with large energy. *MATCH Commun. Math. Comput. Chem.*, 58(1):57–73, 2007.

[56] B. Huo, X. Li, and Y. Shi. Complete solution to a conjecture on the maximal energy of unicyclic graphs. *European J. Combin.*, 32(5):662–673, 2011.

[57] B. Huo, X. Li, and Y. Shi. Complete solution to a problem on the maximal energy of unicyclic bipartite graphs. *Linear Algebra Appl.*, 434(5):1370–1377, 2011.

[58] O. Ivanciuc, T.-S. Balaban, and A. T. Balaban. Design of topological indices. IV. Reciprocal distance matrix, related local vertex invariants and topological indices. *J. Math. Chem.*, 12(1-4):309–318, 1993.

[59] F. Jelen and E. Triesch. Superdominance order and distance of trees with bounded maximum degree. *Discrete Appl. Math.*, 125(2-3):225–233, 2003.

[60] S. Ji, X. Li, and Y. Shi. Extremal matching energy of bicyclic graphs. *MATCH Commun. Math. Comput. Chem.*, 70(2):697–706, 2013.

[61] M. Jooyandeh, D. Kiani, and M. Mirzakhah. Incidence energy of a graph. *MATCH Commun. Math. Comput. Chem.*, 62(3):561–572, 2009.

[62] A. K. Kelmans. The properties of the characteristic polynomial of a graph. In *Cybernetics—in the service of Communism, Vol. 4 (Russian)*, pages 27–41. Izdat. "Ènergija", Moscow, 1967.

[63] L. Kier and L. Hall. *Molecular connectivity in chemistry and drug research*. Academic Press, New York, 1976.

[64] D. Klein, I. Lukovits, and I. Gutman. On the definition of the hyper-Wiener index for cycle-containing structures. *J. Chem. Inf. Comput. Sci.*, 35:50–52, 1995.

[65] D. J. Klein and M. Randić. Resistance distance. *J. Math. Chem.*, 12:81–95, 1993.

[66] J. H. Koolen and V. Moulton. Maximal energy graphs. *Adv. in Appl. Math.*, 26(1):47–52, 2001.

[67] X. Li and Y. Shi. A survey on the Randić index. *MATCH Commun. Math. Comput. Chem.*, 59(1):127–156, 2008.

[68] X. Li, Y. Shi, and I. Gutman. *Graph energy*. Springer, New York, 2012.

[69] X. Li, H. Zhao, and I. Gutman. On the Merrifield-Simmons index of trees. *MATCH Commun. Math. Comput. Chem.*, 54(2):389–402, 2005.

[70] H. Lin and M. Song. On segment sequences and the Wiener index of trees. *MATCH Commun. Math. Comput. Chem.*, 75(1):81–89, 2016.

[71] W. Lin, T. Gao, Q. Chen, and X. Lin. On the minimal ABC index of connected graphs with given degree sequence. *MATCH Commun. Math. Comput. Chem.*, 69(3):571–578, 2013.

[72] J. Liu and B. Liu. A Laplacian-energy-like invariant of a graph. *MATCH Commun. Math. Comput. Chem.*, 59(2):355–372, 2008.

[73] L. Lovász. *Combinatorial problems and exercises*. North-Holland Publishing Co., Amsterdam-New York, 1979.

[74] L. Lovász and J. Pelikán. On the eigenvalues of trees. *Period. Math. Hungar.*, 3:175–182, 1973. Collection of articles dedicated to the memory of Alfréd Rényi, II.

[75] L. Lovász and M. D. Plummer. *Matching theory*, volume 121 of *North-Holland Mathematics Studies*. North-Holland Publishing Co., Amsterdam, 1986.

[76] B. J. McClelland. Properties of the latent roots of a matrix: The estimation of pi-electron energies. *The Journal of Chemical Physics*, 54(2):640–643, 1971.

[77] R. Merrifield and H. Simmons. *Topological methods in chemistry*. Wiley, New York, 1989.

[78] R. E. Merrifield and H. E. Simmons. The structure of molecular topological spaces. *Theor. Chim. Acta*, 55:55–75, 1980.

[79] R. E. Merrifield and H. E. Simmons. Enumeration of structure-sensitive graphical subsets: Calculations. *Proc. Natl. Acad. Sci. USA*, 78:1329–1332, 1981.

[80] R. E. Merrifield and H. E. Simmons. Enumeration of structure-sensitive graphical subsets: Theory. *Proc. Natl. Acad. Sci. USA*, 78:692–695, 1981.

[81] R. E. Merrifield and H. E. Simmons. Topology of bonding in π-electron systems. *Proc. Natl. Acad. Sci. USA*, 82:1–3, 1985.

[82] B. Mohar, D. Babić, and N. Trinajstić. A novel definition of the Wiener index for trees. *J. Chem. Inf. Comput. Sci.*, 33:153–154, 1993.

[83] J. Ou. On extremal unicyclic molecular graphs with prescribed girth and minimal Hosoya index. *J. Math. Chem.*, 42(3):423–432, 2007.

[84] J. Ou. On extremal unicyclic molecular graphs with maximal Hosoya index. *Discrete Appl. Math.*, 157(2):391–397, 2009.

[85] X.-F. Pan and Z.-R. Sun. The (n, m)-graphs of minimum Hosoya index. *MATCH Commun. Math. Comput. Chem.*, 64(3):811–820, 2010.

[86] X.-F. Pan, J.-M. Xu, C. Yang, and M.-J. Zhou. Some graphs with minimum Hosoya index and maximum Merrifield-Simmons index. *MATCH Commun. Math. Comput. Chem.*, 57(1):235–242, 2007.

[87] A. S. Pedersen and P. D. Vestergaard. The number of independent sets in unicyclic graphs. *Discrete Appl. Math.*, 152(1-3):246–256, 2005.

[88] D. Plavšić, S. Nikolić, N. Trinajstić, and Z. Mihalić. On the Harary index for the characterization of chemical graphs. *J. Math. Chem.*, 12(1-4):235–250, 1993.

[89] H. Prodinger and R. F. Tichy. Fibonacci numbers of graphs. *Fibonacci Quart.*, 20(1):16–21, 1982.

[90] M. Randić. Characterization of molecular branching. *J. Amer. Chem. Soc.*, 97(23):6609–6615, 1975.

[91] M. Randić. Novel molecular descriptor for structure-property studies. *Chem. Phys. Lett.*, 211:478–483, 1993.

[92] M. Robbiano and R. Jiménez. Applications of a theorem by Ky Fan in the theory of Laplacian energy of graphs. *MATCH Commun. Math. Comput. Chem.*, 62(3):537–552, 2009.

[93] H. Sachs. Beziehungen zwischen den in einem Graphen enthaltenen Kreisen und seinem charakteristischen Polynom. *Publ. Math. Debrecen*, 11:119–134, 1964.

[94] N. S. Schmuck, S. Wagner, and H. Wang. Greedy trees, caterpillars, and Wiener-type graph invariants. *MATCH Commun. Math. Comput. Chem.*, 68(1):273–292, 2012.

[95] A. J. Schwenk. Almost all trees are cospectral. In *New directions in the theory of graphs (Proc. Third Ann Arbor Conf., Univ. Michigan, Ann Arbor, Mich., 1971)*, pages 275–307. Academic Press, New York, 1973.

[96] R. H. Shi. The average distance of trees. *Systems Sci. Math. Sci.*, 6(1):18–24, 1993.

[97] A. V. Sills and H. Wang. On the maximal Wiener index and related questions. *Discrete Appl. Math.*, 160(10-11):1615–1623, 2012.

[98] W. So and W.-H. Wang. Finding the least element of the ordering of graphs with respect to their matching numbers. *MATCH Commun. Math. Comput. Chem.*, 73(1):225–238, 2015.

[99] D. Spielman. Spectral graph theory. In *Combinatorial scientific computing*, Chapman & Hall/CRC Comput. Sci. Ser., pages 495–524. CRC Press, Boca Raton, FL, 2012.

[100] D. Stevanović. *Spectral Radius of Graphs*. Academic Press, 2014.

[101] L. A. Székely, H. Wang, and T. Wu. The sum of the distances between the leaves of a tree and the 'semi-regular' property. *Discrete Math.*, 311(13):1197–1203, 2011.

[102] R. Tichy and S. Wagner. Extremal Problems for Topological Indices in Combinatorial Chemistry. *J. Comput. Biol.*, 12(7):1004–1013, 2005.

[103] S. Wagner. A class of trees and its Wiener index. *Acta Appl. Math.*, 91(2):119–132, 2006.

[104] S. Wagner. Extremal trees with respect to Hosoya index and Merrifield-Simmons index. *MATCH Commun. Math. Comput. Chem.*, 57(1):221–233, 2007.

[105] S. Wagner. Upper and Lower Bounds for Merrifield-Simmons Index and Hosoya Index. In I. Gutman, B. Furtula, K. C. Das, E. Milovanović, and I. Milovanović, editors, *Bounds in Chemical Graph Theory – Basics*, volume 19 of *Mathematical Chemistry Monographs*, pages 155–187. University of Kragujevac and Faculty of Science Kragujevac, Kragujevac, 2017.

[106] S. Wagner and I. Gutman. Maxima and minima of the Hosoya index and the Merrifield-Simmons index: a survey of results and techniques. *Acta Appl. Math.*, 112(3):323–346, 2010.

[107] S. Wagner, H. Wang, and G. Yu. Molecular graphs and the inverse Wiener index problem. *Discrete Appl. Math.*, 157(7):1544–1554, 2009.

[108] S. Wagner, H. Wang, and X.-D. Zhang. Distance-based graph invariants of trees and the Harary index. *Filomat*, 27(1):41–50, 2013.

[109] B. Wang, C. Ye, and H. Zhao. Extremal unicyclic graphs with respect to Merrifield-Simmons index. *MATCH Commun. Math. Comput. Chem.*, 59(1):203–216, 2008.

[110] H. Wang. The extremal values of the Wiener index of a tree with given degree sequence. *Discrete Appl. Math.*, 156(14):2647–2654, 2008.

[111] H. Wang. Corrigendum: The extremal values of the Wiener index of a tree with given degree sequence. *Discrete Appl. Math.*, 157(18):3754, 2009.

[112] H. Wang. Sums of distances between vertices/leaves in k-ary trees. *Bull. Inst. Combin. Appl.*, 60:62–68, 2010.

[113] H. Wang. The distances between internal vertices and leaves of a tree. *European J. Combin.*, 41:79–99, 2014.

[114] H. Wang and H. Hua. Unicycle graphs with extremal Merrifield-Simmons index. *J. Math. Chem.*, 43(1):202–209, 2008.

[115] H. Wang and G. Yu. All but 49 numbers are Wiener indices of trees. *Acta Appl. Math.*, 92(1):15–20, 2006.

[116] H. Wang and S. Yuan. On the sum of squares of degrees and products of adjacent degrees. *Discrete Math.*, 339(4):1212–1220, 2016.

[117] W. D. Wei. The class $\mathcal{U}(r, s)$ of $(0, 1)$-matrices. *Discrete Math.*, 39(3):301–305, 1982.

[118] H. Wiener. Correlation of heats of isomerization, and differences in heats of vaporization of isomers. *J. Am. Chem. Soc.*, 69:2636–2638, 1947.

[119] H. Wiener. Structural determination of paraffin boiling points. *J. Am. Chem. Soc.*, 69:17–20, 1947.

[120] K. Xu, M. Liu, K. C. Das, I. Gutman, and B. Furtula. A survey on graphs extremal with respect to distance-based topological indices. *MATCH Commun. Math. Comput. Chem.*, 71(3):461–508, 2014.

[121] W. Yan and L. Ye. On the minimal energy of trees with a given diameter. *Appl. Math. Lett.*, 18(9):1046–1052, 2005.

[122] Y. Yang and L. Lu. The Randić index and the diameter of graphs. *Discrete Math.*, 311(14):1333–1343, 2011.

[123] Y. Ye, X.-F. Pan, and H. Liu. Ordering unicyclic graphs with respect to Hosoya indices and Merrifield-Simmons indices. *MATCH Commun. Math. Comput. Chem.*, 59(1):191–202, 2008.

[124] A. Yu and X. Lv. The Merrifield-Simmons indices and Hosoya indices of trees with k pendant vertices. *J. Math. Chem.*, 41(1):33–43, 2007.

[125] P. Yu. An upper bound on the Randić index of trees. *J. Math. Study*, 31:225–230, 1998.

[126] X.-D. Zhang. The signless Laplacian spectral radius of graphs with given degree sequences. *Discrete Appl. Math.*, 157(13):2928–2937, 2009.

[127] X.-D. Zhang, Y. Liu, and M.-X. Han. Maximum Wiener index of trees with given degree sequence. *MATCH Commun. Math. Comput. Chem.*, 64(3):661–682, 2010.

[128] X.-D. Zhang, Q.-Y. Xiang, L.-Q. Xu, and R.-Y. Pan. The Wiener index of trees with given degree sequences. *MATCH Commun. Math. Comput. Chem.*, 60(2):623–644, 2008.

[129] X.-M. Zhang, Y.-Q. Sun, H. Wang, and X.-D. Zhang. On the ABC index of connected graphs with given degree sequences. *J. Math. Chem.*, 56(2):568–582, 2018.

[130] X.-M. Zhang, X.-D. Zhang, R. Bass, and H. Wang. Extremal trees with respect to functions on adjacent vertex degrees. *MATCH Commun. Math. Comput. Chem.*, 78(2):307–322, 2017.

[131] H. Zhao and R. Liu. On the Merrifield-Simmons index of graphs. *MATCH Commun. Math. Comput. Chem.*, 56(3):617–624, 2006.

[132] B. Zhou and I. Gutman. A connection between ordinary and Laplacian spectra of bipartite graphs. *Linear Multilinear Algebra*, 56(3):305–310, 2008.

Index

9781032476032